The 1st International Precision Agriculture Pakistan Conference 2022 (PAPC 2022)—Change the Culture of Agriculture

The 1st International Precision Agriculture Pakistan Conference 2022 (PAPC 2022)—Change the Culture of Agriculture

Editors

Qamar Zaman
Muhammad Jehanzeb Masud Cheema
Muhammad Naveed Tahir
Shoaib Rashid Saleem
Tahir Iqbal
Saddam Hussain
Muhammad Umair
Muhammad Naveed Anjum

MDPI • Basel • Beijing • Wuhan • Barcelona • Belgrade • Manchester • Tokyo • Cluj • Tianjin

MDPI

Editors
Qamar Zaman
PMAS-Arid Agriculture University
Pakistan

Muhammad Jehanzeb Masud Cheema
PMAS-Arid Agriculture University
Pakistan

Muhammad Naveed Tahir
PMAS-Arid Agriculture University
Pakistan

Shoaib Rashid Saleem
PMAS-Arid Agriculture University
Pakistan

Tahir Iqbal
PMAS-Arid Agriculture University
Pakistan

Saddam Hussain
PMAS-Arid Agriculture University
Pakistan

Muhammad Umair
PMAS-Arid Agriculture University
Pakistan

Muhammad Naveed Anjum
PMAS-Arid Agriculture University
Pakistan

Editorial Office
MDPI
St. Alban-Anlage 66
4052 Basel, Switzerland

This is a reprint of Proceedings published online in the open access journal *Environmental Sciences Proceedings* (ISSN 2673-4931) (available at: https://www.mdpi.com/2673-4931/23/1).

For citation purposes, cite each article independently as indicated on the article page online and as indicated below:

LastName, A.A.; LastName, B.B.; LastName, C.C. Article Title. *Journal Name* **Year**, *Volume Number*, Page Range.

ISBN 978-3-0365-7214-7 (Hbk)
ISBN 978-3-0365-7215-4 (PDF)

Contents

About the Editors

Qamar Zaman

Dr. Zaman comes from an average family, whose elders struggled to move to Pakistan from India during the partition in 1947. Dr. Zaman pursued higher education, studying Agricultural Engineering at the University of Agriculture Faisalabad before moving to the UK to complete a PhD in Agriculture Engineering at Newcastle University. Following graduation, Dr. Zaman relocated to the USA, taking up a postdoctoral fellowship at the University of Florida. It was the start of an accomplished academic career that would make him a leading researcher in Precision Agriculture, eventually becoming Chair of the Precision Agriculture Program—a leading research establishment at Dalhousie University of Canada.

Additionally, for twelve-and-a-half years prior to joining the PMAS-AAUR community, he served as Chair Precision Agriculture at the Dalhousie, a globally ranked 200-year-old University in Nova Scotia, Canada. There, he led the development and implementation of new technology for efficiently harvesting wild blueberries in a mechanized way. Prior to joining Dalhousie, Dr. Zaman was involved in a third research venture at Tokyo University of Agriculture and Technology, Japan, under the JSPS program. Currently, Prof. Dr. Qamar uz Zaman is the fourth Vice Chancellor of the second-highest-ranked agricultural institution in Pakistan. He is also author of many books, peer reviewed articles, patents, and other research proceedings. He is also the founder of modern Precision Agricultural practices in Pakistan.

Muhammad Jehanzeb Masud Cheema

Dr. MJM Cheema is a Professor and in charge of the Agricultural Engineering Faculty at PMAS-Arid Agriculture University, Rawalpindi. He also served as the program chair for precision agriculture at the Center for Advance Studies in Agriculture and Food Security. He is a remote sensing expert and has experience using multi-sensor satellite data to manage water resources in data-scarce river basins. He focuses on developing methodologies to efficiently utilize satellite measurements for hydrology and to model conjunctive water use in the transboundary Indus River Basin.

Dr Cheema uses his expertise in the field of precision agriculture and improvements of precision planters, variable rate sprayers and UAVs for agriculture. He has also delivered lectures on the use of remote sensing for water management at various universities in Pakistan and abroad. Currently, he runs a number of nationally and internationally funded projects on water and agriculture and is an author of more than 60 peer reviewed international journal articles and book chapters.

He holds a PhD degree in engineering from the Technical University Delft, the Netherlands, and master's and bachelor's degrees in agricultural engineering with honors from the University of Agriculture, Faisalabad, Pakistan.

Muhammad Naveed Tahir

Dr. Naveed is currently working as an Associate Professor of Agronomy and the director of the Institute of Geoinformation and Earth Observation (IGEO) at PMAS-Arid Agriculture University, Rawalpindi, Pakistan, and also, director of the Pakistan Sub-Center of National Center for International Collaboration Research on Precision Agricultural Aviation Pesticide Spraying Technology, South China Agricultural University, Guangzhou, China. Dr. Naveed graduated from Northwest A&F University, China, in 2012 with a PhD degree. Dr Naveed is also a Project Director (PD) of the federal Government-Funded Project (PSDP) entitled "Pilot Project for Data Driven Smart Decision Platform for Increased Agriculture Productivity (DDSDP)" with a net worth of INR 979.66 million (approx. USD

5 million). For a long time, Dr. Naveed has been engaged in research on remote sensing and Artificial Intelligence of UAVs for precision and digital applications in crop health and disease monitoring fields and UAV-based precision spraying chemical applications for weeds, pests and disease control. He is heading the National and International Research Program with many countries. He is also acting as an Associate Guest editor for a Special Issue in Frontier in Plant Sciences on "Spotlight on Artificial Intelligence for Sustainable Plant Production" and is a Guest Editor for a Special Issue of Remote Sensing, MDPI, "UAV in sustainable Agriculture". He is also a member of many international societies and an editor in internationally renowned journals. Dr Naveed is also supervising PhD and MSc students and has published a number of scientific research papers (40) in SCI/EI Journals.

Shoaib Rashid Saleem

Dr. Shoaib Rashid Saleem is working as Assistant Professor in the Faculty of Agricultural Engineering & Technology and Director, Center for Precision Agriculture (C4PA) at PMAS-Arid Agriculture University, Rawalpindi (PMAS-AAUR). He is actively supervising graduate and undergraduate students. He is also overseeing various development and academic research projects at the C4PA.

Dr. Saleem completed his B.Sc. in Agricultural Engineering from the University of Agriculture, Faisalabad, before moving to Canada for further study. He completed his M.Sc. in Precision Agriculture from Dalhousie University and PhD in Engineering from the University of Guelph. After the completion of his PhD., He worked as postdoctoral fellow at UQAM, Montreal, and as a Hydrologist at Ontario Ministry of Environment, Canada. Dr. Saleem's research focuses on the fundamental understanding and development of state-of-the-art precision agriculture technologies for Eastern Canada's agriculture industry and integrated watershed modelling of agricultural watersheds. Dr. Saleem has worked on the delineation of management zones for site-specific fertilization, remote sensing, groundwater modeling, and hydrological modeling. He has co-supervised 7 graduate students, 20 research associates, and 6 undergraduate summer students over the past 9 years. Dr. Saleem has published 7 peer reviewed journal articles and 19 conference proceedings and conducted 15 industry presentations at various conferences and professional events in relation to precision agriculture, integrated watershed modeling, and climate and land use change impacts on agricultural watersheds.

Tahir Iqbal

Dr. Tahir Iqbal is working as an Associate Professor of Agricultural Engineering at Pir Mehr Ali Shah Arid Agriculture University, Rawalpindi. His research focuses on Farm Mechanization, Precision Agriculture, Renewable Energy, and Resource Conservation. Currently, his work is on Digital and Precision Agriculture Technologies to enable Smart Farms. His group also working on the Conversion of Agricultural Biomass Waste to Energy.

Saddam Hussain

Saddam Hussain, working as a Scientific Officer for Digital Agriculture section at National Center of Industrial Biotechnology, PMAS-Arid Agriculture, Rawalpindi, Pakistan.

Hussain has work experience in the different national and international projects, i.e., PSDP, HEC, PARB and CSIRO. Moreover, he has published 20 peer reviewed articles, 14+ conference proceeding articles and 22+ abstract publications in national and international conferences. He is also a reviewer and editor of many MDPI, Elsevier and Springer journals. Moreover, he is interested to work on precision and digital agriculture, remote sensing, crop modeling and climate change to ensure food security.

Muhammad Umair

Dr. Muhammad Umair is currently working as Associate Professor/Chairman in the department of Energy Systems Engineering, Pir Mehr Ali Sha Arid Agriculture University, Rawalpindi. He completed his master's and Ph.D. degrees in Engineering from Tokyo University of Agriculture & Technology in 2011 and 2014, respectively, where he worked on Solar Adsorption Cooling Systems. His research focus is on Renewable Energies, Solar Cooling, Solar Drying, Bioenergy, etc. He has published more than 30 research papers in national and international journals.

Muhammad Naveed Anjum

Dr. Muhammad Naveed Anjum is an accomplished scholar and expert in the field of Land and Water Conservation Engineering. He currently serves as an Assistant Professor at the PMAS-Arid Agriculture University in Rawalpindi, Pakistan, where he conducts research, teaches courses, and mentors students in the agricultural and water resources engineering disciplines.

Dr. Anjum received his Bachelor's degree in Agricultural Engineering with Honors from the University of Agriculture, Faisalabad, Pakistan, in 2008. He received his Master's degree in Water Resources Engineering from the University of Engineering and Technology, Lahore, Pakistan, in 2010. Dr. Anjum completed his Ph.D. in Hydrology and Water Resources from the University of Chinese Academy of Sciences in 2018. Before pursuing advanced degrees, Dr. Anjum worked as a consulting engineer for Engineering General Consultants in Pakistan, where he gained valuable experience in channel hydraulics, water resource, and irrigation design studies. Since then, he has dedicated his career to advancing the field of hydrology and water resources. His research interests include climate change impact assessment and adaptation on the water sector, land surface hydrology, performance evaluations of satellite-based rainfall products, and projections of future river flows under changing climatic conditions. Dr. Anjum's contributions to his field are extensive, with more than 50 peer reviewed international journal articles. His main research foci have been on the estimation, analysis, and modeling of precipitation, snowpacks, and runoff processes in complex terrain with specific emphasis on the inter-mountain westerlies and monsoon systems.

Preface to "The 1st International Precision Agriculture Pakistan Conference 2022 (PAPC 2022)—Change the Culture of Agriculture"

The first international Precision Agriculture Pakistan Conference (PAPC), in 2022, was held in person at PMAS-Arid Agriculture University Rawalpindi (PMAS-AAUR), Pakistan. The PAPC provided opportunities to local researchers to collaborate with national and international researchers in the field of Digital and Precision Agriculture through oral and poster presentations, exhibits, field demonstrations, as well as discussions and the exchange of information. The world's leading Precision and Digital Agriculture researchers from Canada, China, the USA, Australia and Turkey gave keynote addresses during the PAPC in 2022. The conference introduced new ideas, solutions and research in the agricultural sector, especially considering global climate change, which have been published in this book.

Qamar Zaman, Muhammad Jehanzeb Masud Cheema, Muhammad Naveed Tahir, Shoaib Rashid Saleem, Tahir Iqbal, Saddam Hussain, Muhammad Umair, and Muhammad Naveed Anjum
Editors

environmental sciences proceedings

MDPI

Editorial

Statement of Peer Review †

Qamar Zaman [1,2,*], Muhammad Jehanzeb Masud Cheema [2,3], Muhammad Naveed Tahir [4,5], Shoaib Rashid Saleem [2,6], Tahir Iqbal [2,4], Saddam Hussain [3,7,8], Muhammad Umair [2] and Muhammad Naveed Anjum [2,4]

1 PMAS-Arid Agriculture University, Rawalpindi 46000, Pakistan
2 Faculty of Agricultural Engineering and Technology, PMAS-Arid Agriculture University, Rawalpindi 46000, Pakistan
3 National Center of Industrial Biotechnology, PMAS-Arid Agriculture University Rawalpindi, Rawalpindi 46000, Pakistan
4 Data-Driven Smart Decision Platform, PMAS-Arid Agriculture University, Rawalpindi 46000, Pakistan
5 Department of Agronomy, PMAS-Arid Agriculture University, Rawalpindi 46000, Pakistan
6 Center for Precision Agriculture, PMAS-Arid Agriculture University, Rawalpindi 46000, Pakistan
7 Department of Irrigation and Drainage, University of Agriculture Faisalabad, Faisalabad 36000, Pakistan
8 Department of Biological and Agricultural Engineering, University of California Davis (UC Davis), Davis, CA 95616, USA
* Correspondence: qzaman@dal.ca

† Presented at the International Precision Agriculture Pakistan Conference 2022 (PAPC 2022)—Change the Culture of Agriculture, Rawalpindi, Pakistan, 22–24 September 2022.

In submitting conference proceedings to *Environment Sciences Proceedings*, the volume editors of the proceedings certify to the publisher that all papers published in this volume have been subjected to peer review administered by the volume editors. Reviews were conducted by expert referees to the professional and scientific standards expected of a proceedings journal.

- Type of peer review: double-blind
- Conference submission management system: https://papc.pk/
- Number of submissions sent for review: 57
- Number of submissions accepted: 40
- Acceptance rate: 70%
- Average number of reviews per paper: 2
- Total number of reviewers involved: 38
- Any additional information on the review process: N/A

Citation: Zaman, Q.; Cheema, M.J.M.; Tahir, M.N.; Saleem, S.R.; Iqbal, T.; Hussain, S.; Umair, M.; Anjum, M.N. Statement of Peer Review. *Environ. Sci. Proc.* **2022**, *23*, 41. https://doi.org/10.3390/environsciproc2022023041

Published: 6 February 2023

environmental sciences proceedings

MDPI

Proceeding Paper

Development of Management Zones for Site-Specific Fertilization in Mustard Fields †

Zainab Haroon [1,2,*], Muhammad Jehanzeb Masud Cheema [2], Shoaib Saleem [1,3,4], Muhammad Naveed Anjum [1,2], Muhammad Amin [1,5], Muhammad Naveed Tahir [1,6], Tahir Iqbal [1,4] and Faiza Khan [1,7]

1 Data-Driven Smart Decision Platform, PMAS-Arid Agriculture University, Rawalpindi 46000, Pakistan
2 Department of Land and Water Conservation Engineering, PMAS Arid Agricultural University, Rawalpindi 46000, Pakistan
3 Center for Precision Agriculture, PMAS Arid Agricultural University, Rawalpindi 46000, Pakistan
4 Department of Farm Machinery and Precision Engineering, PMAS Arid Agricultural University, Rawalpindi 46000, Pakistan
5 Institute of Geo-Information & Earth Observation, PMAS-Arid Agriculture University, Rawalpindi 46000, Pakistan
6 Department of Agronomy, PMAS-Arid Agriculture University, Rawalpindi 46000, Pakistan
7 University Institute for Information Technology, PMAS Arid Agricultural University, Rawalpindi 46000, Pakistan
* Correspondence: zainab.haroon@uaar.edu.pk
† Presented at the International Precision Agriculture Pakistan Conference 2022 (PAPC 2022)—Change the Culture of Agriculture, Rawalpindi, Pakistan, 22–24 September 2022.

Abstract: Smart-farming practices can help to overcome the impacts of soil and crop variability on crop yield. This article studies the impact of fertilization treatment on the production of a major oilseed crop, mustard (*Brassica campestris* L.), in an arid region of Pakistan. Soil-nutrient sampling was performed, using 20 × 20 m grids to characterize and quantify the variation of mustard yield and soil nutrients. Based on the results of a significant correlation between yield and soil nutrients, management zones (MZs) were developed that use Variable Rate Fertilization (VRF) to increase crop production and profitability while reducing environmental risk.

Keywords: smart farming; spatial variability; management zones; Variable Rate Fertilization; arid

Citation: Haroon, Z.; Cheema, M.J.M.; Saleem, S.; Anjum, M.N.; Amin, M.; Tahir, M.N.; Iqbal, T.; Khan, F. Development of Management Zones for Site-Specific Fertilization in Mustard Fields. *Environ. Sci. Proc.* **2022**, 23, 1. https://doi.org/10.3390/environsciproc2022023001

Academic Editor: Saddam Hussain

Published: 25 November 2022

1. Introduction

Sustainable agricultural-production systems are necessary to feed the global population, as it is projected to increase by 35% in 2050, which requires a 70–100% increase in food production, and there is an insistent demand to develop this production more efficiently [1].

Mustard (Brassica) is a major oilseed crop grown in Pakistan, but production capacity is limited in Pakistan, despite having huge potential. Therefore, edible-oil import is necessary, which hast cost millions of dollars for many years [2].

Agriculture management in semi-arid and arid regions faces a lot of challenges, because soils are highly variable due to unpredictable rainfall, in terms of both amount and intensity [3]. The precise application of fertilizers can be improved by site-specific nutrient management (SSNM), especially for crops such as rice, maize, wheat, etc. SSNM is a component of precision agriculture (PA), and it helps in improving Nutrient Use Efficiency (NUE) in soil and crops [4].

Therefore, it is necessary to implement efficient techniques to accurately measure variations in soil properties within the field and to draw precise Management Zones (MZs) [5]. To achieve this goal, it is important to explore the crop-production potential, quantify yield gaps, identify limiting variables for enhancing mustard yield, and develop MZs for the precision application of fertilizer.

2. Methodology

A mustard field (1 ha, 33°6′50″ N 73°0′57″ E) was selected during the mustard growing season 2021–2022 at the University Research Farm "Koont", PMAS-Arid Agriculture University Rawalpindi, to investigate the spatial variation effect on mustard yield. The selected field had been under uniform management for the past few decades. Soil variation was studied in the selected field, using grids and soil-nutrient samples, e.g., nitrogen (N), phosphorus (P), and potassium (K), which were recorded with the help of a proximal sensor from each georeferenced grid-point. Yield was measured manually with a hand sickle. The collected data were organized for statistical and geostatistical analysis. The proposed methodology is shown in Figure 1.

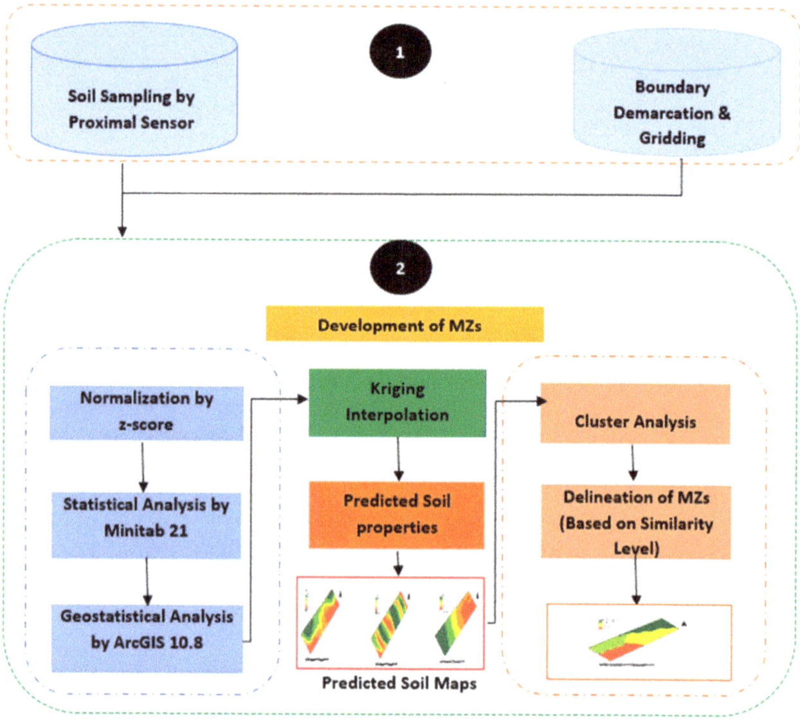

Figure 1. Schematic flow diagram of the proposed methodology.

Descriptive statistics and multivariate geostatistical methods were used to determine the existence and pattern of spatial variability in the selected fields. Descriptive statistics were utilized to calculate the coefficient of variance (CV), which defines the existence of spatial variability in the mustard field, by using Minitab 21 statistical software (Minitab, LLC, State College, PA, USA). Geostatistical analysis was performed using ArcGIS 10.8 software, and semivariogram parameters were estimated, which were used to define the pattern of variability by the kriging interpolation technique. Different models, such as the exponential, circular, spherical, and Gaussian semivariogram models were tested, to find the parameters of the semivariogram nugget, sill, and range [6].

3. Results

3.1. Interpolation of Soil Nutrients and Mustard Yield

There was significant spatial variation in the mustard field, as evidenced by the kriged interpolated maps of sol nutrients and mustard yield, requiring the establishment of MZs for site-specific nutrient management. Higher values of nutrients were mostly observed on the south sid, and lower values on the central and north sides of the field, as indicated in Figure 2.

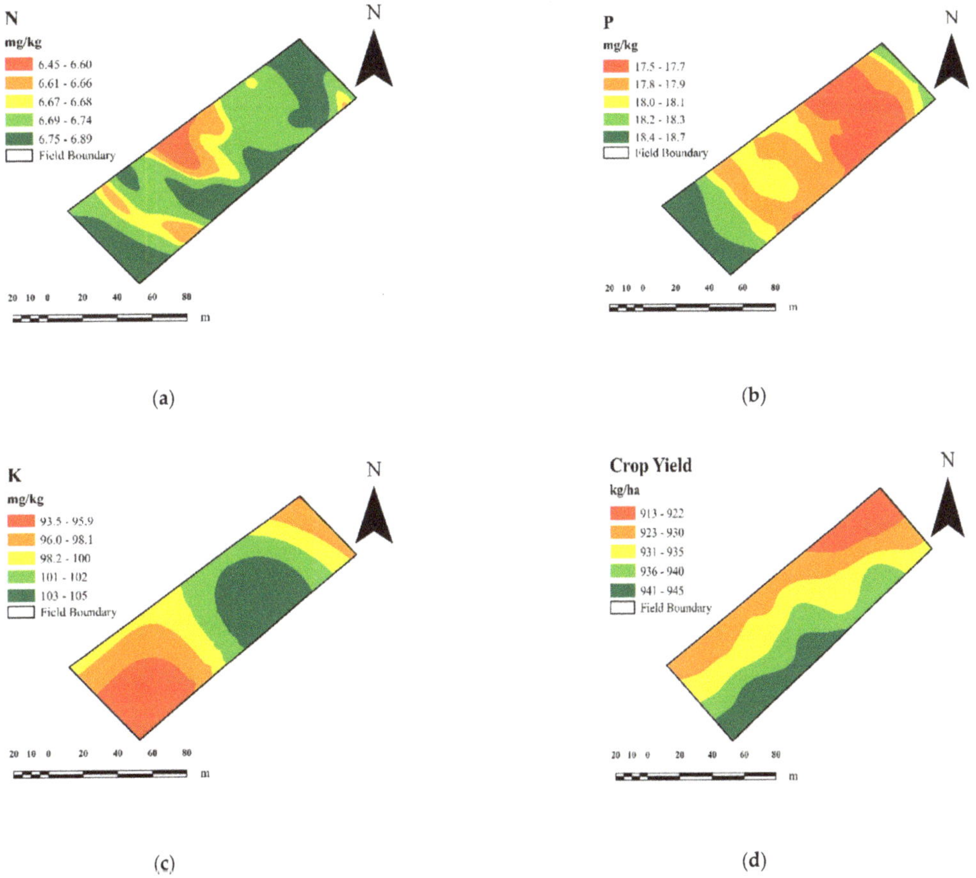

Figure 2. Kriged maps of soil nutrients and crop yield for the mustard crop, (**a**) N; (**b**) P; (**c**) K; (**d**) Crop Yield.

The visualization of the maps revealed that variation in soil properties caused substantial variation in yield. The high-yielding values were observed on the north-south side of the field, and low values were on the northwest side of the field. The results of interpolated maps suggested moderate-to-high variation in nutrients and mustard yield.

Delineation of MZs for Mustard Crop

The clustered soil nutrients and mustard-yield data were imported into ArcGIS 10.8 software to analyze and develop the MZs. The developed MZs, on the basis of clusters, were classified into four productivity zones: very poor, poor, medium, and good for mustard, as shown in Figure 3.

The delineated MZs showed different productivity levels within a field, which required variable-rate fertilizer-application to improve the mustard yield. A higher productivity level was observed on the southeast side of the field. The southwest and the middle of the field had average productivity, whereas the northern side had extremely low productivity. Developed MZs can be used for site-specific soil- and crop-sampling, significantly reducing farm-production costs while protecting the environment from the harmful effects of chemical fertilizer. The authors in [7] summarized how the developed MZs can reduce the over- and under-application of fertilization for wild blueberries, which reduces the cost of production of the farm and increases the yield. Up to 40% of agrochemicals was saved by the accurate appliication of fertilizer according to the developed MZs [8].

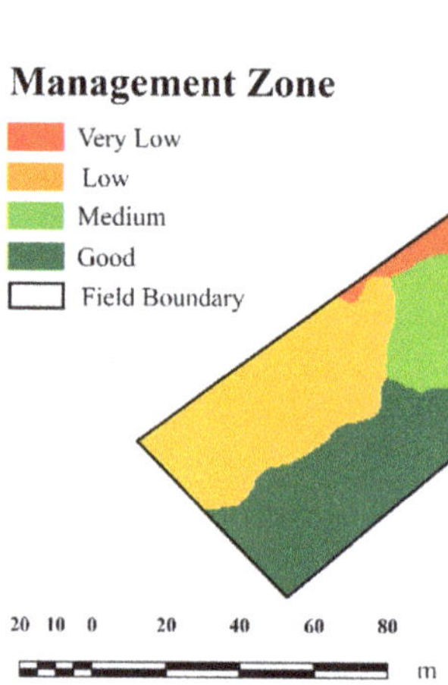

Figure 3. Delineated MZs for precise application of fertilizer in the mustard crop.

4. Conclusions

Soil-sampling planning and management of mustard crops is significantly assisted by an understanding of soil parameters. Cluster analysis could provide a method for planning and organizing the spatial variability of soil parameters and crop production within individual fields. Input-use efficiency, production costs, and environmental advantages could all improve with the implementation of MZs-based variable-rate fertilizers. To better understand the effect of climate change on crop-production variability and to determine whether or not MZs have the opportunity to conduct site-specific nutrient management, it would be useful for future research to correlate soil attributes with climate change.

Author Contributions: Conceptualization, Z.H. and M.J.M.C.; methodology, Z.H.; software, M.J.M.C.; validation, Z.H., S.S. and M.A.; formal analysis, Z.H.; investigation, S.S.; resources, M.J.M.C.; data curation, Z.H. and F.K.; writing—original draft preparation, Z.H., S.S, M.N.A. and F.K.; writing—review and editing, T.I. and Z.H.; visualization, T.I. and S.S.; supervision, M.J.M.C.; project administration, M.N.T.; funding acquisition, Z.H. All authors have read and agreed to the published version of the manuscript.

Funding: The study is a part of a pilot project for the Data-Driven Smart Decision Platform for Increased Agriculture Productivity (DDSDP) PSDP-funded project No. 332.

Institutional Review Board Statement: Not applicable.

Informed Consent Statement: Not applicable.

Data Availability Statement: The data can be available on request.

Acknowledgments: The authors would like to thank Faculty of Agricultural Engineering and Technology, PMAS-Arid Agriculture University Rawalpindi, Pakistan and DDSDP project.

Conflicts of Interest: The authors declare no conflict of interest.

References

1. Van Wart, J.; Kersebaum, K.C.; Peng, S.; Milner, M.; Cassman, K.G. Estimating crop yield potential at regional to national scales. *Field Crop. Res.* **2013**, *143*, 34–43. [CrossRef]
2. Jat, R.A.; Wani, S.P.; Sahrawat, K.L. Conservation agriculture in the semi-arid tropics: Prospects and problems. *Adv. Agron.* **2012**, *117*, 191–273.
3. Waqas, M.S.; Cheema, M.J.M.; Hussain, S.; Ullah, M.K.; Iqbal, M.M. Delayed irrigation: An approach to enhance crop water productivity and to investigate its effects on potato yield and growth parameters. *Agric. Water Manag.* **2021**, *245*, 106576. [CrossRef]
4. Sarkar, D.; Meena, V.S.; Haldar, A.; Rakshit, A. Site-specific nutrient management (SSNM): A unique approach towards maintaining soil health. In *Adaptive Soil Management: From Theory to Practices*; Springer: Singapore, 2017; pp. 69–88.
5. Peralta, N.R.; Costa, J.L. Delineation of management zones with soil apparent electrical conductivity to improve nutrient management. *Comput. Electron. Agric.* **2013**, *99*, 218–226. [CrossRef]
6. Shukla, A.K.; Behera, S.K.; Lenka, N.K.; Tiwari, P.K.; Prakash, C.; Malik, R.S.; Chaudhary, S.K. Spatial variability of soil micronutrients in the intensively cultivated Trans-Gangetic Plains of India. *Soil Tillage Res.* **2016**, *163*, 282–289. [CrossRef]
7. Farooque, A.A.; Zaman, Q.U.; Schumann, A.W.; Madani, A.; Percival, D.C. Delineating management zones for site specific fertilization in wild blueberry fields. *Appl. Eng. Agric.* **2012**, *28*, 57–70. [CrossRef]
8. Khan, H.; Farooque, A.A.; Acharya, B.; Abbas, F.; Esau, T.J.; Zaman, Q.U. Delineation of Management Zones for Site-Specific Information about Soil Fertility Characteristics through Proximal Sensing of Potato Fields. *Agronomy* **2020**, *10*, 1854. [CrossRef]

environmental sciences proceedings

MDPI

Proceeding Paper

Managing the Food Security Nexus under Climate Change: Recent Advances in Precision Agriculture Practices in Pakistan †

Sidra Javed [1,*], Asif Sardar [1], Aneela Afzal [1,2], Arshad Mahmood Malik [1,3], Muhammad Jehanzeb Masud Cheema [1,4] and Sehrish Kanwal [5]

1 National Center of Industrial Biotechnology, PMAS Arid Agriculture University Rawalpindi, Rawalpindi 46000, Pakistan
2 Department of Agriculture Extension, PMAS Arid Agriculture University Rawalpindi, Rawalpindi 46000, Pakistan
3 Department of Economics, PMAS Arid Agriculture University Rawalpindi, Rawalpindi 46000, Pakistan
4 Faculty of Engineering and Technology, PMAS Arid Agriculture University Rawalpindi, Rawalpindi 46000, Pakistan
5 Data-Driven Smart Decision, Platform, PMAS Arid Agriculture University Rawalpindi, Rawalpindi 46000, Pakistan
* Correspondence: sidrajaved161@gmail.com
† Presented at the International Precision Agriculture Pakistan Conference 2022 (PAPC 2022)—Change the Culture of Agriculture, Rawalpindi, Pakistan, 22–24 September 2022.

Abstract: Climate change poses a significant challenge to food security. Several traditional approaches are available to prevent food insecurity in climate change scenarios, such as modifications in crop practices, altering field management systems, and climate-smart farming measures. Recently, precision agriculture practices are becoming a more popular way of farming that might help to overcome climate risk and reduce food insecurity. Precision agriculture is a tool for monitoring the food supply chain and managing both the amount and quality of agricultural products. Thus, this work seeks to synthesize and discuss the research related to the importance of a precision agriculture system in preventing food insecurity in the context of a developing countries such as Pakistan. The study is based on the narrative literature review.

Keywords: food security; precision agriculture; adaptive strategies; climate change

Citation: Javed, S.; Sardar, A.; Afzal, A.; Malik, A.M.; Cheema, M.J.M.; Kanwal, S. Managing the Food Security Nexus under Climate Change: Recent Advances in Precision Agriculture Practices in Pakistan. *Environ. Sci. Proc.* **2022**, *23*, 2. https://doi.org/10.3390/environsciproc2022023002

Academic Editor: Shoaib Rashid Saleem

Published: 25 November 2022

1. Introduction

Food insecurity through climate change can occur directly through the changes in ecological parameters and indirectly through disturbances in the economic growth and reduction in agricultural products [1]. The global climate change influences on food production and distribution are more severe in developing countries than in developed countries. Many adaptations and mitigation techniques, such as improving seed quality, modifying cropping practices, introducing new varieties, and improving irrigation techniques, are available to improve food production during the climatic shift [2]. However, these techniques could not improve much yield. There is a need for an integrated approach that improves the yield without compromising the environment. Thus, precision agriculture is the sole solution to this problem [3]. Precision agriculture is a collection of technologies that integrate sensors, information systems, advanced equipment, and well established management to maximize productivity through reducing variation and uncertainty in agricultural systems [4]. Adapting agricultural inputs within a field to site-specific conditions allows for a more efficient allocation of energy, which helps to preserve ecological integrity while improving the food production system [5]. The precision agriculture strategies not only improve the economic condition of the farmers but also reduce the negative impacts of extensive agriculture practices on the environment. Moreover, this study fulfills the literature gap of relationship of the tradition agriculture practices with precision agriculture. This study examines the potential of available mitigation and adaptation approaches to prevent food security in climate change situations. Furthermore, it also explores the importance of precision agriculture technologies in elimination of food insecurity in Pakistan.

2. Methodology

This study reviewed the secondary published data from different reliable sources to identify the influence of climate change on food security. To critically analyses the situation in developing countries such as Pakistan, the authors carried out a desk review of the relevant reports, projects, papers, and web-based publications related to food availability, accessibility, and usage in developing countries and Pakistan.

3. Global Food Security and Climate Change

Climate change has an impact on both food security and the lives of farmers involved in agricultural systems. It also influenced the value chains of the agricultural systems. According to the report, about 70% of the total global population lives in some countries/territories experiencing a food crisis, including Afghanistan, the Democratic Republic of the Congo, Sudan, Ethiopia, northern Nigeria, Yemen, Pakistan, the Syrian Arab Republic, South Sudan, and Haiti [6]. The global prevalence of hunger is alarmingly high. According to the report, food insecurity levels in 2021 were higher than any previously recorded, with about 193 million people in need of immediate assistance across 53 countries/territories. An increase of approximately 40 million people was reported in 2020 [6].

The main drivers of food insecurity in developing countries are societal conflicts, economic shocks, and weather extremes. According to Figure 1, 139.1 million people in 53 countries are facing food crises caused by conflicts, 23.5 million by weather extremes, and 30.2 million by economic shock in developing countries [7]. Pakistan already has a high rate of food insecurity, and the recent floods have just exacerbated the situation in the country. According to global food hunger index, Pakistan lies at 92 of a total of 116 countries [6].

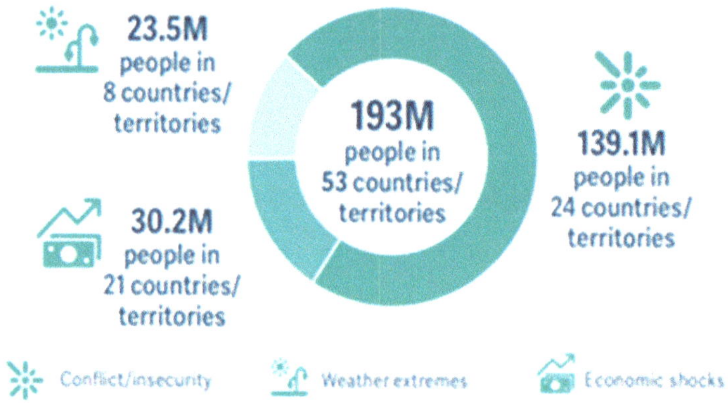

Figure 1. Global food crisis [6].

4. Mitigation and Adaptation Measures for Sustainable Agriculture

To deal with food insecurity caused by climate change, various mitigation and adaptation measures are available, such as usage of precision agriculture technologies [8], crop varieties that can withstand climate change, improved agronomic techniques, water and irrigation conservation measures, a diverse agricultural economy, and agroforestry, which is stated by the studies. However, the implications of these techniques are limited to specific areas and conditions.

5. Recent Advances in Precision Agriculture Technologies

Scientists from all over the globe have been trying to find different ways to solve the problem of food insecurity. The ideal alternative to reduce the risk of food insecurities is "precise agriculture". Precision agriculture integrates sensors, information systems, and improved equipment (see for details in Figure 2) and guides the management to maximize agricultural system productivity without damaging the environment.

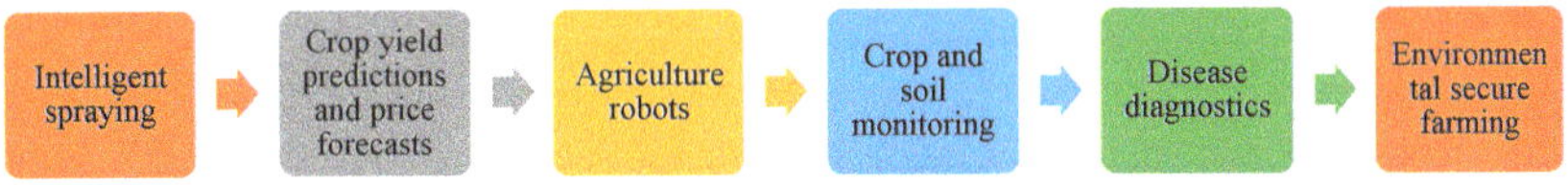

Figure 2. Role of precision agriculture in sustainable agriculture production.

The large-scale application of precision agriculture not only revolutionizes the agriculture systems of developing countries, but it also prevents the further degradation of the environment. Adoption of precision agriculture can provide the following benefits:

- Precision agriculture provides more efficient use of resources, which improves the long-term viability of food production without compromising environmental standards.
- Precision agriculture allows for the management of both the quantity and quality of agricultural products by paying close attention to the whole food production chain.
- Precision agricultural technologies, according to current research, can reduce production costs by up to 20% while increasing productivity and protecting the environment.
- To meet the challenges of fluctuating weather and rising food demand, precision agriculture seeks to invigorate the combination of farm improvement and climate susceptibility [9].

Climate change poses an ever-increasing danger to the billions of people around the globe whose livelihoods depend on agriculture production. These changes could slow a country's economic growth, halting the progress it has made over many years. Precision agriculture technologies are already used in developed countries. Developing countries are also attempting to apply these technologies in the true sense by capitalizing on their technologies, seeking new sources of funding, promoting climate-smart methods, and allowing the institution to operate [1].

The precision agriculture management system adoption is suggested for developments in both conventional agricultural technology (such as tractors, crop genetics, equipment, herbicides, and fertilizers) and information technology (such as GNSS, sensors, computer processing capabilities, telematics, and variable rate technology). The use of precision agriculture technology in crop management systems is a great addition to any conventional farm [4]. With precision agriculture, farmers may apply inputs in a more targeted manner, leading to increased yield, lower cost, more consistent output, and fewer negative environmental impacts [10].

6. Recommendations

Food security is a challenge for developing countries. There is a need for policy guidance regarding the modern agriculture system in developing countries such as Pakistan. Moreover, the government should promote the interdisciplinary efforts of academic research with industrial linkages to promote the adoption of precision agriculture at the farm level.

7. Conclusions

Food security is a challenge for developing countries, as food insecurity is rising with climate shifts. The catastrophic events now frequently occur due to the climate shift, further exacerbating the condition. Several mitigation and adoption approaches are available to overcome the food security issues, especially those caused by climatic changes, i.e., the introduction of modified crop varieties, implementation of soil conservation methods, efficient irrigation management techniques, renewable energy, etc. Thus, precision agriculture technologies could provide a better solution for sustainable crop yield while improving the quality to meet the rising demand of the population without compromising the environment.

Author Contributions: Conceptualization, S.J. and A.S.; methodology, S.J. and A.S.; investigation, S.J., A.S., A.A., A.M.M., M.J.M.C. and S.K.; writing—original draft preparation, S.J., A.S., A.A., A.M.M., M.J.M.C. and S.K.; writing—review and editing and S.J., A.S., A.A., A.M.M., M.J.M.C. and S.K.; supervision A.S., A.A. and M.J.M.C.; project administration, A.A. and M.J.M.C. All authors have read and agreed to the published version of the manuscript.

Funding: The study is a part of a PSDP-funded project (no. 321—"Establishment of National Center of Industrial Biotechnology for Pilot Manufacturing of Bioproducts Using Synthetic Biology and Metabolic Engineering Technologies at PMAS-Arid Agriculture University Rawalpindi").

Institutional Review Board Statement: Not applicable.

Informed Consent Statement: Not applicable.

Data Availability Statement: Not applicable.

Acknowledgments: We thank the organizing committee of the 1st Annual International Precision Agriculture Pakistan Conference, held on 22–24 September 2022. The authors also gratefully acknowledge PSDP funded project no. 321—"Establishment of National Center of Industrial Biotechnology for Pilot Manufacturing of Bioproducts Using Synthetic Biology and Metabolic Engineering Technologies at PMAS-Arid Agriculture University Rawalpindi", executed through the Higher Education Commission Islamabad, Pakistan.

Conflicts of Interest: The authors declare no conflict of interest.

References

1. Espitia, A.; Evenett, S.; Rocha, N.; Ruta, M. Widespread Food Insecurity Is Not Inevitable: Avoid Escalating Food Export Curbs. Available online: https://cepr.org/system/files/publication-files/172987-global_economic_consequences_of_the_war_in_ukraine_sanctions_supply_chains_and_sustainability.pdf#page=130 (accessed on 1 October 2022).
2. WRI. WRI Annual Report 2020-21 | World Resources Institute. 2021. Available online: https://www.wri.org/annualreport/2020-21 (accessed on 29 September 2022).
3. Hussain, S.; Amin, A.; Mubeen, M.; Khaliq, T.; Shahid, M.; Hammad, H.M.; Sultana, S.R.; Awais, M.; Murtaza, B.; Amjad, M.; et al. Climate Smart Agriculture (CSA) Technologies. In *Building Climate Resilience in Agriculture*; Springer: Cham, Switzerland, 2021; pp. 319–338. [CrossRef]
4. Say, S.M.; Keskin, M.; Sehri, M.; Emre Sekerli, Y. Adoption of Precision Agriculture Technologies in Developed and Developing Countries. In Proceedings of the International Science and Technology Conference (ISTEC), Berlin, Germany, 17–19 July 2017; Available online: https://www.researchgate.net/publication/320908156_Adoption_of_Precision_Agriculture_Technologies_in_Developed_and_Developing_Countries (accessed on 29 September 2022).
5. Bramley, R.; Trengove, S. Precision agriculture in Australia: Present status and recent developments. *Eng. Agrícola* **2013**, *33*, 575–588. [CrossRef]
6. FAO. FSIN Food Security Information Network. 2022. Available online: https://www.fao.org/3/cb9997en/cb9997en.pdf (accessed on 29 October 2022).
7. Karbekova, A.B.; Samieva, K.T. Imbalances in Food Security of the World Countries as a Problem of Sustainable Agricultural Development. In *Sustainable Agriculture*; Springer: Singapore, 2022; pp. 191–197. [CrossRef]
8. Shahzad, A.; Ullah, S.; Dar, A.A.; Sardar, M.F.; Mehmood, T.; Tufail, M.A.; Shakoor, A.; Haris, M. Nexus on climate change: Agriculture and possible solution to cope future climate change stresses. *Environ. Sci. Pollut. Res. Int.* **2021**, *28*, 14211–14232. [CrossRef] [PubMed]
9. Waqas, M.S.; Cheema, M.J.M.; Hussain, S.; Ullah, M.K.; Iqbal, M.M. Delayed irrigation: An approach to enhance crop water productivity and to investigate its effects on potato yield and growth parameters. *Agric. Water Manag.* **2021**, *245*, 106576. [CrossRef]
10. Kiani, A.K.; Sardar, A.; Khan, W.U.; He, Y.; Bilgic, A.; Kuslu, Y.; Raja, M.A.Z. Role of Agricultural Diversification in Improving Resilience to Climate Change: An Empirical Analysis with Gaussian Paradigm. *Sustainability* **2021**, *13*, 9539. [CrossRef]

environmental sciences proceedings

MDPI

Proceeding Paper

Role of 5G and 6G Technology in Precision Agriculture †

Syed Muhammad Zaigham Abbas Naqvi [1,2,3,4], Shoaib Rashid Saleem [2,5,*], Muhammad Naveed Tahir [1,2], Shixin Li [3], Saddam Hussain [5,6,7], Syed Ijaz Ul Haq [1,2,8] and Muhammad Awais [2]

1 Department of Agronomy, PMAS-Arid Agriculture University, Rawalpindi 46300, Pakistan
2 Data Driven Smart Decision Platform (DDSDP) Project, PMAS-Arid Agriculture University, Rawalpindi 46300, Pakistan
3 College of Mechanical and Electrical Engineering, Henan Agricultural University, Zhengzhou 450002, China
4 Henan International Joint Laboratory of Laser Technology in Agriculture Sciences, Zhengzhou 450002, China
5 Faculty of Agricultural Engineering and Technology, PMAS-Arid Agriculture University, Rawalpindi 46300, Pakistan
6 Department of Irrigation and Drainage, University of Agriculture Faisalabad, Faisalabad 38000, Pakistan
7 National Centre of Industrial Biotechnology, PMAS-Arid Agriculture University, Rawalpindi 46300, Pakistan
8 College of Agricultural Engineering and Food Science, Shandong University of Technology, Zibo 255022, China
* Correspondence: shoaibrashid@uaar.edu.pk; Tel.: +92-302-6109556

† Presented at the International Precision Agriculture Pakistan Conference 2022 (PAPC 2022)—Change the Culture of Agriculture, Rawalpindi, Pakistan, 22–24 September 2022.

Abstract: Smart farming and precision agriculture rely on the different components of IoT, such as sensors, drones, and robotic devices. IoT in agriculture is the network of interconnected devices that corresponds in real time, simultaneously, to gather, analyze, and transfer the data, which, ultimately, generate a decision to be taken by the farmer. The availability of the 4G/3G does not support the precision practices in real time due to the bandwidth, connectivity, and the speed of data-transfer issues. Further, 5G technology in the agricultural sector has put its greater influence in real-time monitoring, unmanned aerial vehicles, virtual consultation and predictive maintenance, artificially intelligent robotics, and data analytics and cloud repositories. Conclusively, the speed, connectivity, scalability and processing power, and limitations can be overcome with the availability of 5G structures.

Keywords: 5G; precision agriculture; 6G; real-time monitoring

Citation: Naqvi, S.M.Z.A.; Saleem, S.R.; Tahir, M.N.; Li, S.; Hussain, S.; Ul Haq, S.I.; Awais, M. Role of 5G and 6G Technology in Precision Agriculture. *Environ. Sci. Proc.* **2022**, 23, 3. https://doi.org/10.3390/environsciproc2022023003

Academic Editor: Muhammad Jehanzeb Masud Cheema

Published: 25 November 2022

1. Introduction

Agriculture is the main source of survival for all living beings on Earth. The systemic production of crop plants via the integration of modern technological advancements can mitigate the upcoming disasters of food security and hunger risk. The technological integration of smart agriculture, precision agriculture, 5G, and IoT can boost farm practices at the level of automation that can transform a decision-support system to optimize operations [1]. The evolutionary period of agriculture has reached the level of digital agriculture, named as "Agriculture 4.0" or digital agriculture (Figure 1) [2]. Developing countries are lacking in the adaptation of digital agriculture due to their limited sources.

Cellular communications and internet connectivity for modern times require fast services for heavy files of data transfer with high speed. These targets are now made possible to achieve by the modern fifth-generation (5G) networks, with the characteristics of achieving universal connectivity, minimum latency, and, specifically, with the highest rate of data transfer [3]. Although the technology is moving towards the next phase of the novel state-of-the-art sixth-generation (6G) network, currently, the global system is running on 5G. The IoT is completely dependent on the availability of interconnected devices that can collect, analyze, and transfer data in real time. The use of IoT in agriculture is advancing rapidly with the advent of 5G because IoT devices can perform long-range operations rapidly and reliably. Thus, 5G technology in the agricultural sector has put its greater influence in real-time monitoring, unmanned aerial vehicles, virtual consultation and predictive maintenance, artificially intelligent robotics, and data analytics and cloud repositories (Figure 2) [4].

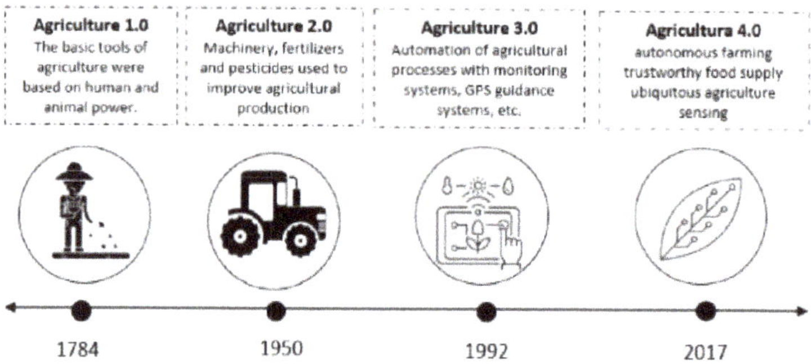

Figure 1. The evolutionary phases of agriculture [2].

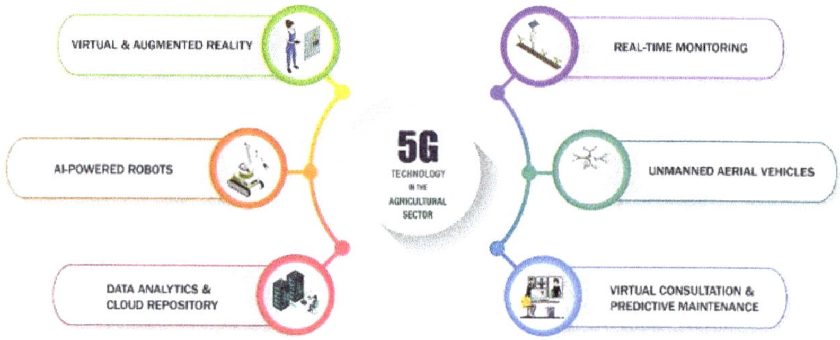

Figure 2. 5G technology in the agriculture sector [4], copyright permission: License Number: 5411150959041; License date: 17 October 2022; Licensed Content Publisher: Elsevier.

2. Application of 5G in Agriculture

The conventional methods of agriculture are being adapted toward the new era of smart farming, which enables the production of higher yields of crops and less human interference. The components of smart farming, including sensors, collect large amounts of data and transfer them remotely in real time. This 5G technology fulfils the demand of smart farming by predicting and preventing the crops from disease damage using mobile phones. Not only this but 5G in agriculture has enormous applications, such as unmanned aerial vehicles, monitoring in real time, climate change mitigations, visual monitoring and predictive measures, artificially empowered robotics, and data analytics and cloud repository. Automated driving systems, such as autosteering tractors, deep learning, and cloud-based mobile applications, connected with 5G can further alleviate the yield outputs by many folds with maximized efficiency from farmers. Full exploration of 5G technology will define the overall productivity of precision agriculture over time [5] (Table 1).

Table 1. Exploration of 5G and 6G technology in agricultural systems.

Agricultural Tasks	Why 5G and 6G?
Real-time monitoring	Energy efficiency Device density & data volume Ultralow latency Reliability
UAVs	High data rate Ultralow latency Energy consumption Security
Augmented reality and virtual reality	High data rate Low latency
Virtual consultation and predictive maintenance	High data rate Ultralow latency Reliability
AI-driven robots	High data rate Low latency Reliability Low energy consumption Security
Data analytics and repository	High data rate Ultralow latency Reliability Security

3. Related Work

Although developed countries are working on 5G and IoT devices in agriculture collectively to obtain their massive benefits for quantity and quality enhancements of crops, researchers are still working in the developing regions of the world to obtain the maximum output from IoT devices used in the field. The requirements of smart farming and precision agriculture were analyzed to integrate them with 5G technology for better production [4]. A protype of sustainable greenhouse was designed and implemented in Nigeria to assess the performance of IoT devices interconnected with 5G for an information-processing framework. The framework developed the artificial methods for reasoning and imaging that ultimately improved the high quality of cultivation management and ease of access to the greenhouse technology [6].

The integration of artificial intelligence and a wireless communication system was performed in the MERLIN project in India that made efficient use of 5G/6G and IoT devices. The project was run to develop a technology solution, which used cloud-based databases, namely supervised and semi-supervised, to obtain data and further process them for decision making. The decision-support system can be automated, which can improve the production and yield of the crop plants and make efficient use of remote sensors and IoT integrated with 5G/6G [7]. Smart farming was further enhanced by performing agricultural management and focusing on the development of smart systems using 5G mobile networks. The smart system was able to transfer data at a high speed, up to 20 Gbps. This system can link the enormous amount of IoT devices in a square kilometer [8].

Implementation of 5G/6G technology in Colombia boosted the use of IoT applications and increased the demand for digital agriculture. The scenarios were developed for 5G/6G usage with various bands of frequencies, their applications, and use in the farmland activities. The rural-area-based scenarios in Colombia were developed in real time for data acquisition by integrating 5G with the interconnected devices of Mobile edge and robotics that produced significant results for crop development in the remote areas of villages [8,9].

4. Challenges and Future Perspective of 5G

IoT-based devices are interconnected and wirelessly connected in a smart farm system through the availability of 4G/3G/NB-IoT. Smart devices are available for the transfer, collection of the data, precise analyses, estimation, and production of decision-supported results in farmlands. However, the networking system that interconnects these sensory systems possesses some limitations that bar the full use of precision farming technologies. The most significant limitation is the operational region because developing countries, such as Pakistan, lack the coverage of wireless networks in remote areas. City areas face the problem of connectivity, quality of services, channel conditions, and fluctuations in data-transfer rates in 4G. Another issue is the battery longevity, which is drained due to the multiple antennas and transmitters attached to the devices. IoT devices, such as robotics, drones, and field sensors, are operated remotely and remain for a longer time in the field, so cannot be sustained with the varying connectivity in 4G networks.

The connectivity issues, most importantly, hinder the advancements of precision agriculture in developing and underdeveloped regions of the world. Currently, the sensors on drones acquire images from the field and then are transferred to sprayer drones. If 5G technology is available, it can perform real-time image collection and spot spraying can be performed without lagging. Speed, connectivity, scalability, and processing power are all limitations that can be overcome with the availability of 5G structures. The global word is the old concept, while 5G is the future of the world that will soon be running every single industry for human beings.

Author Contributions: Conceptualization, S.M.Z.A.N., S.R.S. and S.H.; methodology, S.I.U.H., M.A. and S.H.; writing—original draft preparation, S.M.Z.A.N.; writing—review and editing, S.R.S., S.H. and S.L.; supervision, M.N.T., S.R.S., S.H. and S.L.; funding acquisition, M.N.T. All authors have read and agreed to the published version of the manuscript.

Funding: The study is a part of a pilot project for Data-Driven Smart Decision Platform for Increased Agriculture Productivity (DDSDP) PSDP-funded project No. 332.

Institutional Review Board Statement: Not applicable.

Informed Consent Statement: Not applicable.

Data Availability Statement: Not applicable.

Acknowledgments: The author is also very thankful to the Higher Education Commission, Islamabad, Pakistan, for supporting the study under the funded project No. 332.

Conflicts of Interest: The authors declare no conflict of interest.

References

1. Arrubla-Hoyos, W.; Ojeda-Beltrán, A.; Solano-Barliza, A.; Rambauth-Ibarra, G.; Barrios-Ulloa, A.; Cama-Pinto, D.; Arrabal-Campos, F.M.; Martínez-Lao, J.A.; Cama-Pinto, A.; Manzano-Agugliaro, F. Precision Agriculture and Sensor Systems Applications in Colombia through 5G Networks. *Sensors* **2022**, *22*, 7295. [CrossRef] [PubMed]
2. DayioČLu, M.A.; Turker, U. Digital Transformation for Sustainable Future-Agriculture 4.0: A review. *J. Agric. Sci.* **2021**, *27*, 373–399. [CrossRef]
3. Panwar, N.; Sharma, S.; Singh, A.K. A survey on 5G: The next generation of mobile communication. *Phys. Commun.* **2016**, *18*, 64–84. [CrossRef]
4. Tang, Y.; Dananjayan, S.; Hou, C.; Guo, Q.; Luo, S.; He, Y. A survey on the 5G network and its impact on agriculture: Challenges and opportunities. *Comput. Electron. Agric.* **2021**, *180*, 105895. [CrossRef]
5. Torky, M.; Hassanein, A.E. Integrating blockchain and the internet of things in precision agriculture: Analysis, opportunities, and challenges. *Comput. Electron. Agric.* **2020**, *178*, 105476. [CrossRef]
6. Guo, X. Application of agricultural IoT technology based on 5 G network and FPGA. *Microprocess. Microsyst.* **2021**, *80*, 103597. [CrossRef]
7. Sharma, A. MERLIN: Smart Framework for Agriculture in India. In Proceedings of the 2021 International Conference on Information and Communication Technology Convergence (ICTC), Jeju Island, Korea, 20–22 October 2021; pp. 453–457.

8. Said Mohamed, E.; Belal, A.A.; Kotb Abd-Elmabod, S.; El-Shirbeny, M.A.; Gad, A.; Zahran, M.B. Smart farming for improving agricultural management. *Egypt. J. Remote. Sens. Space Sci.* **2021**, *24*, 971–981. [CrossRef]
9. Valecce, G.; Strazzella, S.; Grieco, L.A. On the Interplay between 5G, Mobile Edge Computing and Robotics in Smart Agriculture Scenarios. In Proceedings of the Ad-Hoc, Mobile, and Wireless Networks, Cham, Switzerland, 1 October 2019; pp. 549–559.

environmental sciences proceedings

MDPI

Proceeding Paper

Vertical Farming—Current Practices and Its Future †

Syed Muhammad Zaigham Abbas Naqvi [1,2,3,4], Shoaib Rashid Saleem [2,5,*], Muhammad Naveed Tahir [1,2], Saddam Hussain [5,6,7], Syed Ijaz Ul Haq [1,2,8], Muhammad Awais [2] and Salman Qamar [2]

1 Department of Agronomy, PMAS-Arid Agriculture University, Rawalpindi 46300, Pakistan
2 Data Driven Smart Decision Platform (DDSDP) Project, PMAS-Arid Agriculture University, Rawalpindi 46300, Pakistan
3 College of Mechanical and Electrical Engineering, Henan Agricultural University, Zhengzhou 450002, China
4 Henan International Joint Laboratory of Laser Technology in Agriculture Sciences, Zhengzhou 450002, China
5 Faculty of Agricultural Engineering and Technology, PMAS-Arid Agriculture University, Rawalpindi 46300, Pakistan
6 Department of Irrigation and Drainage, University of Agriculture Faisalabad, Faisalabad 38000, Pakistan
7 National Centre of Industrial Biotechnology, PMAS-Arid Agriculture University, Rawalpindi 46300, Pakistan
8 College of Agricultural Engineering and Food Science, Shandong University of Technology, Zibo 255022, China
* Correspondence: shoaibrashid@uaar.edu.pk; Tel.: +92-302-6109556
† Presented at the International Precision Agriculture Pakistan Conference 2022 (PAPC 2022)—Change the Culture of Agriculture, Rawalpindi, Pakistan, 22–24 September 2022.

Abstract: The depletion of usable agricultural lands has brought up a scenario of vertical farming. This type of farming is mostly considered soil-less farming in the vertical direction. Three of the commonly used soil-less ways for vertical farming include hydroponic, aeroponic, and aquaponic. Although it is not very popular in developing countries, investment has been made by many European counties and efforts to use vertical farming as a commercial product are on the path to success. Food security issues can be addressed through this farming type as well.

Keywords: soil-less agriculture; vertical farming; food security

Citation: Naqvi, S.M.Z.A.; Saleem, S.R.; Tahir, M.N.; Hussain, S.; Ul Haq, S.I.; Awais, M.; Qamar, S. Vertical Farming—Current Practices and Its Future. *Environ. Sci. Proc.* **2022**, *23*, 4. https://doi.org/10.3390/environsciproc2022023004

Academic Editor: Muhammad Jehanzeb Masud Cheema

Published: 25 November 2022

1. Introduction

The growth in population around the globe is demanding living space and land to grow food and food products. The fertility of agricultural land is decreasing day by day due to the excessive use of synthetic chemicals such as fertilizers, pesticides, insecticides, etc. Researchers are working on soil-less and less space-consuming solutions to grow plants. One of the solutions consists of vertical farming in which plants can be grown in a space that is not of particular advantage, such as used shipping containers, buildings on damaged lands, or warehouses [1]. Although this kind of farming does not require a large space to develop, the environmental parameters need to be strictly monitored. The controlled environment for vertical farming needs to be fulfilled with monitored nutrients, temperature, and light.

Vertical farming practices do not require soil particularly, so plantation can go in an upwards direction in the form of layers and sometimes reach up to multistory levels (Figure 1) [2]. There are several types of vertical farming but three of them are commonly used soil-less methods including hydroponic, aeroponic, and aquaponic. Hydroponic farming is the technique of growing plants in water. The nutrients are continuously monitored and are provided only in calculated amounts. Aeroponic farming systems do not even need a water medium to grow plants. This system only uses a mist or spray of nutrients in the root growth chamber. Aeroponic farming can even save up to 90% of the water as compared to hydroponic. Aquaponic systems are only used for fast-growing crops because this system combines hydroponic farming and fish farming at the same time. Fish growth is reliant on nutrients and fish farming waste, rich in nutrients, is used as a recycling component for plant growth which results in economically viable and efficient plant growth [3].

Figure 1. The vertical agriculture growing in several layers and less space [2].

2. Vertical Farming System

2.1. Crop Selection

There is no hard and fast rule about the selection of crops to grow in vertical farming because the environmental conditions are controlled artificially. However, to choose the best crop that can grow under artificial light and less space is the best tactic for this style of farming. Long, heightened plants such as avocado, olive, banana, and other trees are limited to growing indoors; however, other controlled techniques such tunnel or greenhouse farming can be applied for tree farming. Vertical farming, especially hydroponic, offers the production of more than three dozen various vegetables indoors. Vegetables such as lettuce, spinach, leafy greens, strawberries, cucumber, herbs, tomatoes, etc., are now the crops most grown under vertical farming. However, other crops such as maize, wheat, biofuel, and herbal medicinal crops can also be grown under a vertical cropping system [4].

2.2. Environment Control System

Heating, cooling, and proper air ventilation are very important in vertical farming to keep the best air quality, save energy, and keep a consistent moisture content. Reusable energy resources such as solar energy, geothermal equipment, and ground water usage can help to maintain the low energy cost and keep the environment clean. Control of the indoor environment such as dehumidification can be maintained by natural ventilation, the use of desiccant, and the condensation process. Water chambers to circulate water can be used to lower the temperature in the vertical chamber.

2.3. Waste Management System

In an aquaculture system, the yearly biowaste produced is estimated as 527 t. Biowaste is comprised of dead leaves, parts of stems, fibrous roots, dead fruits, and vegetables and can be converted into organic fertilizers, biofuels, and liquid organic nutrients. Waste water treatment can also make water in reusable conditions using a SlurryCarb machine [5].

2.4. Smart Devices

The vertical farming system operating with automated services without human interventions needs to be operated by sensors and smart devices. The sensory system collects information about the environment, crop health, nutrients, and water requirements and forwards it to interconnected devices. The data are analyzed and a quick decision is made through the decision support system. The requirements of plants are fulfilled by an automatic system without human involvement in agricultural practices [6].

3. Related Work

Although developed countries have many resources to fulfill the feeding needs of their community, European countries and private organizations are working on the efficient use of free spaces such as restaurants, metro stations, supermarkets, and in-store farming. Agricool, a company based in France, is working on container vertical farming for strawberry cultivations [7]. An Estonian company named Click and Grow is working on a novel idea of appliance farms, also called small indoor gardens (Figure 2) [8,9].

Figure 2. An appliance farm for indoor farming [9]; Copyright permission: License Number: 5423060894672; License date: 6 November 2022; Licensed Content Publisher: Elsevier.

An in-store aeroponic farming setup was tested in Italy under the joint venture of Agricooltur and Auchan retailers. The same setup was organized in a supermarket in Luxemburg. In Madrid, in-farm vertical farming was adopted for the purpose of the fresh supply of lettuce to the customers [10]. Another German retail chain, Metro, collaborated with Infarm to grow vertical agriculture within the store and provide fresh herbs and vegetables to the customers [11].

4. Benefits, Challenges and Future Perspective of Vertical Farming

Feeding the world in the 21st century is not an easy task due to several issues of uncontrolled population growth, uneven distribution of resources, depletion of natural resources, overburdened urbanization, loss of fertile soils, etc. Vertical farming can be seen as an opportunity to mitigate food security risks. Vertical farming can provide continuous crop production and is much more efficient; one acre of vertical farming can cover the food production of 30 acres of farming on the land. Due to the controlled environmental conditions, there is less chance of diseases and insects/pest attacks which can eliminate the chance of chemical use during farming practices. Many environmental factors such as hail, flood, drought, etc., that cause crop failure are also eliminated due to controlled environmental conditions. Moreover, vertical farming helps to reduce carbon emissions generated during agronomic practices and reduce water losses by 70% [12].

Although this cropping system can deal with food security issues, it comes with some challenges to deal with. The main challenge to deal with is the cost–benefit analysis. The land and building costs may vary from region to region but cost in urban areas does not allow vertical farming in big cities. The use of energy and operational cost to maintain and control the internal environmental conditions of a farming setup can challenge cost efficiency. A limited number of crops can only be grown through vertical farming and also require pollination by hand during crop growth. However, the system lacks some economic efficiencies, but in the future, to avoid food scarcity, vertical farming will become a trend and the reuse of necessary building structures will reduce the cost of the system.

Author Contributions: Conceptualization, S.M.Z.A.N., S.R.S. and S.H.; methodology, S.I.U.H. and M.A.; writing—original draft preparation, S.M.Z.A.N.; writing—review and editing, S.R.S., S.H. and S.Q.; supervision, M.N.T. and S.R.S.; funding acquisition, M.N.T. All authors have read and agreed to the published version of the manuscript.

Funding: This research was funded by the pilot project for data-driven smart decision platform for increased agriculture productivity, grant number PSDP-332.

Institutional Review Board Statement: Not applicable.

Informed Consent Statement: Not applicable.

Data Availability Statement: Not applicable.

Acknowledgments: The author is also very thankful to the Higher Education Commission, Islamabad, Pakistan for supporting the study under the project entitled "Pilot Project for Data Driven Smart Decision Platform for Increased Agriculture Productivity".

Conflicts of Interest: The authors declare no conflict of interest.

References

1. Benke, K.; Tomkins, B. Future food-production systems: Vertical farming and controlled-environment agriculture. *Sustain. Sci. Pract. Policy* **2017**, *13*, 13–26. [CrossRef]
2. Birkby, J. Vertical farming. *ATTRA Sustain.* **2016**, *2*, 1–12.
3. AlShrouf, A. Hydroponics, aeroponic and aquaponic as compared with conventional farming. *Am. Acad. Sci. Res. J. Eng. Technol. Sci.* **2017**, *27*, 247–255.
4. Elouafi, I.; Shahid, M.A.; Begmuratov, A.; Hirich, A. The Contribution of Alternative Crops to Food Security in Marginal Environments. In *Emerging Research in Alternative Crops*; Hirich, A., Choukr-Allah, R., Ragab, R., Eds.; Springer International Publishing: Cham, Switzerland, 2020; pp. 1–23.
5. Cicekli, M.; Barlas, N.T. Transformation of today greenhouses into high technology vertical farming systems for metropolitan regions. *J. Environ. Prot. Ecol.* **2014**, *15*, 1779–1785.
6. Kalantari, F.; Mohd Tahir, O.; Mahmoudi Lahijani, A.; Kalantari, S. A Review of Vertical Farming Technology: A Guide for Implementation of Building Integrated Agriculture in Cities. *Adv. Eng. Forum* **2017**, *24*, 76–91. [CrossRef]
7. Dillet, R. Agricool Raises Another $28 Million to Grow Fruits in Containers. Available online: https://techcrunch.com/2018/12/03/agricool-raises-another-28-million-to-grow-fruits-in-containers/ (accessed on 6 November 2022).
8. Grow, C. Click & Grow Raises $11 Million in Investment. Available online: https://asia.clickandgrow.com/blogs/news/click-grow-raises-11-million-in-investment (accessed on 6 November 2022).
9. Butturini, M.; Marcelis, L.F.M. Chapter 4—Vertical farming in Europe: Present status and outlook. In *Plant Factory*, 2nd ed.; Kozai, T., Niu, G., Takagaki, M., Eds.; Academic Press: Wageningen, The Netherlands, 2020.
10. Souissi, M. A Greenhouse in an Auchan Hypermarket. Available online: https://www.internationalsupermarketnews.com/a-greenhouse-in-an-auchan-hypermarket/ (accessed on 6 November 2022).
11. Maxwell, M. Metro Cash & Carry Trials In-Store Farms. Available online: https://www.fruitnet.com/eurofruit/metro-cash-and-carry-trials-in-store-farms/168307.article (accessed on 6 November 2022).
12. Kulak, M.; Graves, A.; Chatterton, J. Reducing greenhouse gas emissions with urban agriculture: A Life Cycle Assessment perspective. *Landsc. Urban Plan.* **2013**, *111*, 68–78. [CrossRef]

environmental sciences proceedings

MDPI

Proceeding Paper

Determinants of Farmers' Climate-Smart Agriculture Adoption in the Photohar Region †

Asif Sardar [1,*], Aneela Afzal [1,2], Arshad Mahmood Malik [1,3], Muhammad Jehanzeb Masud Cheema [1,4], Saddam Hussain [5,6] and Sidra Javed [1]

1 National Center of Industrial Biotechnology (NCIB), PMAS Arid Agriculture University Rawalpindi, Rawalpindi 44000, Pakistan
2 Department of Agriculture Extension, PMAS Arid Agriculture University, Rawalpindi 44000, Pakistan
3 Department of Economics, Faculty of Social Sciences, PMAS Arid Agriculture University, Rawalpindi 44000, Pakistan
4 Department of Agriculture Engineering and Technology, PMAS Arid Agriculture University Rawalpindi, Rawalpindi 44000, Pakistan
5 Department of Irrigation and Drainage, University of Agriculture Faisalabad, Faisalabad 38000, Pakistan
6 Department of Biological and Agricultural Engineering, University of California Davis (UC Davis), Davis, CA 95616, USA
* Correspondence: asifsardar.erzurum@gmail.com
† Presented at the International Precision Agriculture Pakistan Conference 2022 (PAPC 2022)—Change the Culture of Agriculture, Rawalpindi, Pakistan, 22–24 September 2022.

Abstract: Long-term changes in climate variability, vulnerability, a rise in average temperature, and changes in precipitation patterns threaten crop productivity, food security, and the livelihoods of people across the globe. Lower crop yields with higher dependence on food imports, global economic shocks, and climate change exacerbate more challenges to food security, specifically in developing countries. Therefore, adaptation to climate change is necessary to promote farmers' sustainable livelihoods and mitigate carbon emissions. The adoption of Climate-Smart Agriculture (CSA) practices can potentially help reduce greenhouse gas (GHG) emissions without compromising agricultural production. To confront the challenges regarding the adoption of CSA practices, this study reviews the relevant literature and suggests policy recommendations on how socioeconomic determinants and considerations affect sustainable agriculture development systems. It focuses on the optimization problem of a farmer as a social planner, in which a farmer seeks to maximize his welfare objectives now and in the future. Farmers' choices of CSA adoption, as well as their adaptive capacity and adaptation constraints, are discussed. This study has implications for policymakers in terms of raising the frequency of adopters through innovations and policy design.

Keywords: climate vulnerability; adaptive capacity; crop yield; adaptation; farm income

Citation: Sardar, A.; Afzal, A.; Malik, A.M.; Cheema, M.J.M.; Hussain, S.; Javed, S. Determinants of Farmers' Climate-Smart Agriculture Adoption in the Photohar Region. *Environ. Sci. Proc.* **2022**, *23*, 5. https://doi.org/10.3390/environsciproc2022023005

Academic Editor: Muhammad Umair

Published: 25 November 2022

1. Introduction

Climate change poses an environmental threat to the whole world. Developing countries, such as Pakistan, are more vulnerable to climate-induced events. Every year, Pakistan allocates limited resources for climate change mitigation. Climate change, poverty, and hunger are major issues in Pakistan, where the majority of the population is engaged in the agriculture sector. Between 19 and 29% of total GHG emissions are emitted from the agriculture sector in Pakistan. The Potohar zone lies from about 32.5° N to 34.0° N in latitude and from about 72° E to 74° E in longitude. Barley, bajra, wheat, maize, and groundnuts are the main crops produced in this region. Around two-thirds of the climate of Pakistan is arid, mountainous, and humid. In the Potohar region, agriculture production depends upon the monsoon rainfall. The Photohar region's climate is changing, as evidenced by changes in rainfall intensity and pattern. This may be due to natural processes, but anthropogenic activities are also responsible for this climate change [1]. As a result, Climate-Smart Agriculture (CSA) practices are essential for reducing emissions. CSA is defined as an integrated approach that is used to manage livestock, fisheries, forests, landscapes, and cropland [2]. It addresses the issue of food security by minimizing the impacts of climate change. CSA increases productivity and resilience and reduces emissions [3,4]. CSA is an integrated approach that is comprised of tree planting, crop rotation, soil, and water management measures, and the adoption of smart measures for agriculture production. The adoption of CSA depends on socioeconomic determinants, the adaptive capacity of the farmers, and the socioeconomic profiles of the region. Climate change has impacted livelihoods, particularly the livelihoods of those who rely on the agriculture sector [5]. Therefore, the adoption of CSA is necessary to mitigate the impact of climate change and raise food security. The purpose of this study was to investigate which factors that affect the determinants of CSA adoption and how farmers' adaptive capacity can help them to adopt.

2. Theoretical Framework

The conceptual framework for the adoption of CSA depends upon climate change and its vulnerability, adaptation process, and outcome. In this study, we assumed that a rational farmer wants to adopt CSA practices to maximize their expected profits in the following ways (see Figure 1). This framework demonstrates that a farmer's adoption of CSA is characterized by their adaptive capacity, which is determined by their socioeconomic characteristics.

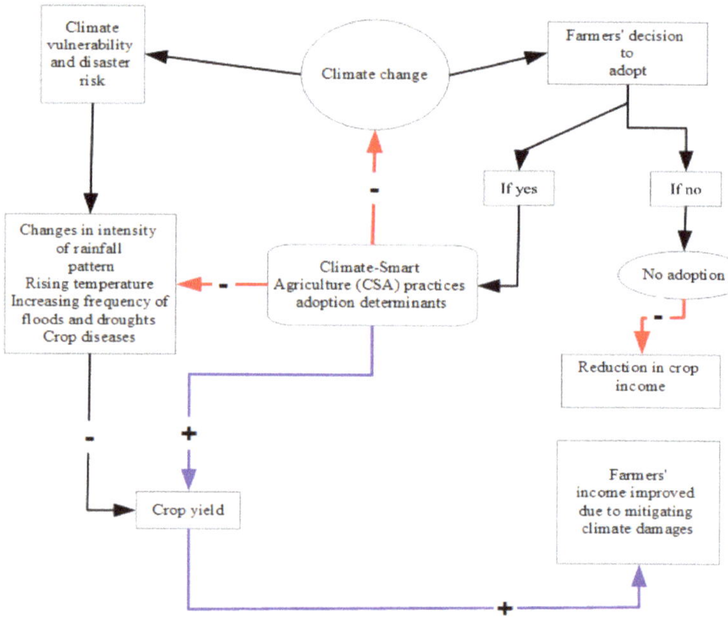

Figure 1. Theoretical framework for the determinants of CSA adoption [3].

3. Methodology

We reviewed published research papers and reports. We filtered out research articles that were published in local and international peer-reviewed journals. We assessed articles on climate-smart agriculture adoption, taking into account the drivers characterized by farmers' socioeconomic determinants that may have influenced adaptation and their intention to adopt CSA practices. Environmental and socioeconomic factors and key constraints determining the adoption of CSA practices were explored. Therefore, the graphical, empirical, and descriptive literature was used to fill the research gap, which can be helpful for policymakers in designing a framework to encourage the adoption of innovative agriculture technologies at the farm level.

4. Results and Discussion

When farmers, as social planners, decide to improve their households' welfare, they must consider different kinds of constraints, such as political, social, behavioral, biophysical, and institutional constraints, to understand the heterogeneous effects of climate change. This expands opportunities to invest in CSA research and development while also assisting farmers by providing farm-level resources and technologies [6]. Selecting a suitable and reasonable model is critical. It depends on socioeconomic and environmental factors. Therefore, in calculating socially optimal solutions, we must take all four elements of food security into account to obtain robust estimates through modeling [7]. According to the study, the socioeconomic characteristics of the farmers are important drivers of CSA adoption (see Figure 2).

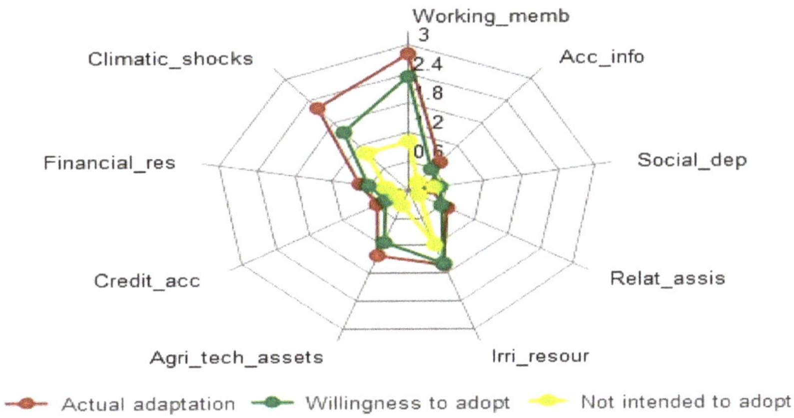

Figure 2. Main socioeconomic determinants of CSA adoption [8].

Therefore, to obtain a reliable estimate of CSA adoption, a farmer must take all relevant factors into account for the modeling, which helps with weighing the multiple objectives to drive a reliable solution. These measures would be helpful in obtaining the best outcome from the socially optimal possible solutions to the farmers' problems to maximize crop revenue. The study also found that in order to increase the frequency of adoption, consumer-driven barriers to CSA adoption must be reduced while taking into account political, behavioral, and social contexts to encourage farmers. Social organizations and networking can assist the farming community in understanding the damage caused by climate change and how to mitigate it. Overall, adopters remained in benefits and gained more farm income, ranging from 45% to 48% more per hectare (ha) than the farmers who did not adopt. Therefore, adopters are better off in terms of gains in crop yield and farm income than the rest of the farmers [3].

5. Conclusions

We (people) are responsible for climate change as 'donors' due to our anthropogenic activities. We are also 'receptors' of these impacts. Climate change impacts are determined by countries' coping capacities, which are based on socioeconomic determinants. Adoption is required to mitigate rising environmental damage in agriculture production, environmental sustainability, and food security. This study found that technological and financial resource availability, higher crop yield and input use efficiency, coordination among the various levels of institutions and departments, social networking, and improvements in social capital for enhancing farmers' adaptive capacity are important determinants of CSA adoption. Overall, adopters of CSA practices remain better off than nonadopters. Effective participation is required to implement CSA measures at the farm level. To increase the number of adopters, socioeconomic policy instruments, such as defining property rights, encouraging social bonding, and networking, providing subsidies for CSA technology, and imposing Pigouvian taxes to discourage nonadopting farmers from engaging in polluting activities at the farm level, are suggested.

Author Contributions: Conceptualization, A.S. and A.A.; methodology, S.J. and A.S.; investigation, A.S., A.A., A.M.M., M.J.M.C. and S.H.; writing—original draft preparation, S.J., A.S., A.A., A.M.M., M.J.M.C. and S.H.; writing—review and editing and A.S., A.A., A.M.M., M.J.M.C. and S.H; supervision A.S., A.A. and M.J.M.C.; project administration, A.A. and M.J.M.C. All authors have read and agreed to the published version of the manuscript.

Funding: The study is a part of PSDP-funded project No. 321.

Institutional Review Board Statement: Not applicable.

Informed Consent Statement: Not applicable.

Data Availability Statement: Not applicable.

Acknowledgments: We thank the organizing committee of the First International Precision Agriculture Pakistan Conference 2022 as well as the Environmental Sciences Proceedings. The authors also gratefully acknowledge PSDP-funded project No. 321, "Establishment of National Center of Industrial Biotechnology for Pilot Manufacturing of Bioproducts Using Synthetic Biology and Metabolic Engineering Technologies at PMAS-Arid Agriculture University Rawalpindi", executed through the Higher Education Commission, Islamabad, Pakistan.

Conflicts of Interest: The authors declare no conflict of interest.

References

1. Jat, H.S.; Choudhary, M.; Datta, A.; Yadav, A.K.; Meena, M.D.; Devi, R.; Gathala, M.K.; Jat, M.L.; McDonald, A.; Sharma, P.C. Temporal changes in soil microbial properties and nutrient dynamics under climate smart agriculture practices. *Soil Tillage Res.* **2020**, *199*, 104595. [CrossRef] [PubMed]
2. Amin, A.; Nasim, W.; Mubeen, M.; Sarwar, S.; Urich, P.; Ahmad, A.; Wajid, A.; Khaliq, T.; Rasul, F.; Hammad, H.M.; et al. Regional climate assessment of precipitation and temperature in Southern Punjab (Pakistan) using SimCLIM climate model for different temporal scales. *Theor. Appl. Climatol.* **2018**, *131*, 121–131. [CrossRef]
3. Sardar, A.; Kiani, A.K.; Kuslu, Y. Does adoption of climate-smart agriculture (CSA) practices improve farmers' crop income? Assessing the determinants and its impacts in Punjab province, Pakistan. *Environ. Dev. Sustain.* **2021**, *23*, 10119–10140. [CrossRef]
4. Mizik, T. Climate-Smart Agriculture on Small-Scale Farms: A Systematic Literature Review. *Agronomy* **2021**, *11*, 1096. [CrossRef]
5. Taneja, G.; Pal, B.D.; Joshi, P.K.; Aggarwal, P.; Tyagi, N.K. Farmers' Preferences for Climate-Smart Agriculture—An Assessment in the Indo-Gangetic Plain. In *Climate Smart Agriculture in South Asia*; Springer: Singapore, 2019; pp. 91–111.
6. Taylor, M. Climate-smart agriculture: What is it good for? *J. Peasant. Stud.* **2017**, *45*, 89–107. [CrossRef]
7. Martey, E.; Etwire, P.M.; Abdoulaye, T. Welfare impacts of climate-smart agriculture in Ghana: Does row planting and drought-tolerant maize varieties matter? *Land Use Policy* **2020**, *95*, 104622. [CrossRef]
8. Sardar, A.; Kiani, K.A.; Kuslu, Y. An Assessment of Willingness for Adoption of Climate-Smart Agriculture (CSA) Practices through the Farmers' Adaptive Capacity Determinants. *Yüzüncü Yil Üniversitesi J. Agric. Sci.* **2019**, *29*, 781–791.

environmental sciences proceedings

MDPI

Proceeding Paper

Deep Learning-Based Approach for Weed Detection in Potato Crops †

Faiza Khan [1,2,*], Noureen Zafar [1,2], Muhammad Naveed Tahir [1,3], Muhammad Aqib [2,4], Shoaib Saleem [1,5] and Zainab Haroon [1,5]

1 Data-Driven Smart Decision Platform, PMAS-Arid Agriculture University, Rawalpindi 46000, Pakistan
2 University Institute for Information Technology, PMAS Arid Agricultural University, Rawalpindi 46000, Pakistan
3 Department of Agronomy, PMAS-Arid Agriculture University, Rawalpindi 46000, Pakistan
4 Establishment of National Center for Industrial Biotechnology (NCIB), PMAS-Arid Agriculture University, Rawalpindi 46000, Pakistan
5 Faculty of Agricultural Engineering and Technology, PMAS Arid Agricultural University, Rawalpindi 46000, Pakistan
* Correspondence: faiza_khan@uaar.edu.pk

† Presented at the 1st International Precision Agriculture Pakistan Conference 2022 (PAPC 2022)—Change the Culture of Agriculture, Rawalpindi, Pakistan, 22–24 September 2022.

Abstract: The digital revolution is transforming agriculture by applying artificial intelligence (AI) techniques. Potato (*Solanum tuberosum* L.) is one of the most important food crops which is susceptible to different varieties of weeds which not only lower its yield but also affect crop quality. Artificial Intelligence and Computer Vision (CV) techniques have been proven to be state-of-the-art in terms of addressing various agricultural problems. In this study, a dataset of five different potato weeds was collected in different environments and under different climatic conditions such as sunny, cloudy, partly cloudy, and at different times of the day on a weekly basis. For weeds-detection purposes, the Tiny-YOLOv4 model was trained on the collected potato weeds dataset. The proposed model obtained 49.4% mAP value by calculating the IoU. The model trained with high prediction accuracy will later be used as part of a site-specific spraying system to apply agrochemicals for weed management in potato crops.

Keywords: deep learning; object detection; YOLO; weed detection; digital agriculture

Citation: Khan, F.; Zafar, N.; Tahir, M.N.; Aqib, M.; Saleem, S.; Haroon, Z. Deep Learning-Based Approach for Weed Detection in Potato Crops. *Environ. Sci. Proc.* **2022**, *23*, 6. https://doi.org/10.3390/environsciproc2022023006

Academic Editor: Muhammad Jehanzeb Masud Cheema

Published: 28 November 2022

1. Introduction

Artificial Intelligence (AI) and Deep Learning (DL) is the best choice for researchers in the field of agriculture for weed detection. DL-based Convolutional Neural Networks (CNNs) embedded with Graphics Processing Unit (GPUs) have resulted in innovations in the agriculture sector [1]. Recently, it played a vital role in automating the process of harvesting by fast and more accurate detection of weeds in the real-time environment. The state-of-the-art deep CNNs can extract complex and useful features from the input images which results in significant detection of weeds [2], diseases [3], pests [4], plant nutrients [5], etc.

Potato (*Solanum tuberosum* L.) is one of the most important food crops for over a billion people worldwide. According to a study in [6], 37% of the potato crop is damaged due to weeds. Piyazi booti (*Asphodelus tenuifolius* L.), Canada thistle (*Cirsium arvense* L.), jungli gajjar (*Parthenium hysterophorous* L.), bathu (*Chenopodium album* L.), and billi booti (*Anagallis arvensis* L.) were the most common weeds that grow in the potato field, especially in Potohar region. Tiny-YOLOv4 model is capable of detecting and classifying/descriminating these weeds in the real-time environment with high prediction accuracy.

The image dataset of potato weeds is not available publicly for training and validation of the model. The findings from this study will later be utilized in the real-time detection and spot-specific sprayer system for the management of the potato field. The first objective of this study was to collect the dataset locally from the Potohar region for training and validating the model. The second was to detect and classify the most common types of weeds in a potato field.

2. Methodology

Various DL and Computer Vision (CV) techniques are being used in the agriculture sector which has become a hot research topic nowadays. The proposed methodology for weed detection and classification is shown in Figure 1. First, the image dataset of five potato weeds was collected through a Logitech camera. Next, image processing was performed; in this step, the blurred and noisy images were discarded. Then, data annotation was performed, and the processed images were labeled through Yolo_mark and labellmg tools. The next step was to split the dataset into training and testing categories for feeding into the selected Tiny-YOLOv4 model. The best trained weights of the model with maximum accuracy were selected for the real-time detection of potato weeds.

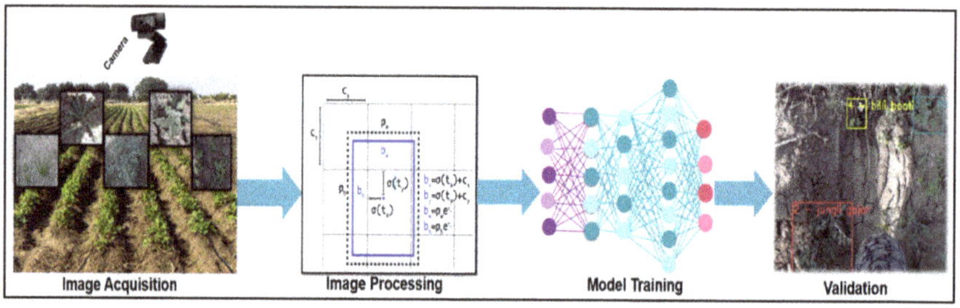

Figure 1. Proposed methodology diagram.

3. Results

3.1. Evaluation Indicators

Precision, Recall, F1-Score, and Mean Average Precision (mAP) were employed to evaluate the prediction model performance in this work.

An alternative term for precision is "positive predictive value". This is calculated by dividing the True Positive (TP) by the sum of TP and False Negative (FN) i.e., (TP + FN) as shown in Equation (1). TP means that the weed belongs to class 1 and the model predicted correctly, that it is from class 1. The FN means that the weed is from class 1, but the model predicted that it belongs to no class. The range for the best value for precision is 1.0, and 0.0 for the worst condition.

$$P = \frac{TP}{TP + FN} \tag{1}$$

where P = Precision.

The Recall is also called "sensitivity" or "true positive rate" and it is the ratio of TP and the sum of TP and FN as shown in Equation (2). FP here means that the weed belongs to class 1 but the model classifies it into class 2. Its value varies from 0 to 1.

$$\text{Recall} = \frac{TP}{TP + FP} \tag{2}$$

The harmonic mean of recall and precision is known as the F1-Score where the best value is 1.0 and 0.0 is the worst. It is calculated by using Equation (3).

$$F1 = 2 \times \frac{\text{precision} * \text{Recall}}{\text{precision} + \text{recall}} \tag{3}$$

The sum of the precision and recall of the detected bounding boxes produces the mean average precision (mAP). Equation (4) provides a formula for calculating mAP, where AP is the average precision of each class.

$$\mathrm{mAP} = 1/\mathrm{N}\sum_{\mathrm{i}=1}^{\mathrm{N}} \mathrm{AP} \tag{4}$$

3.2. Experimental Results

The Tiny-YOLOv4 model was tested on unseen images with the 416 × 416 image resolution to make the models consistent with the training dataset. The experimental results of the Tiny-YOLOv4 model are presented in Figure 2. The red color line represents the mAP, the blue color shows the loss, and the green color represents the iteration count. The model was trained on the 10,000 iterations, 16 subdivisions, and with the 416 × 416 image resolution. The Mish was used as the activation function.

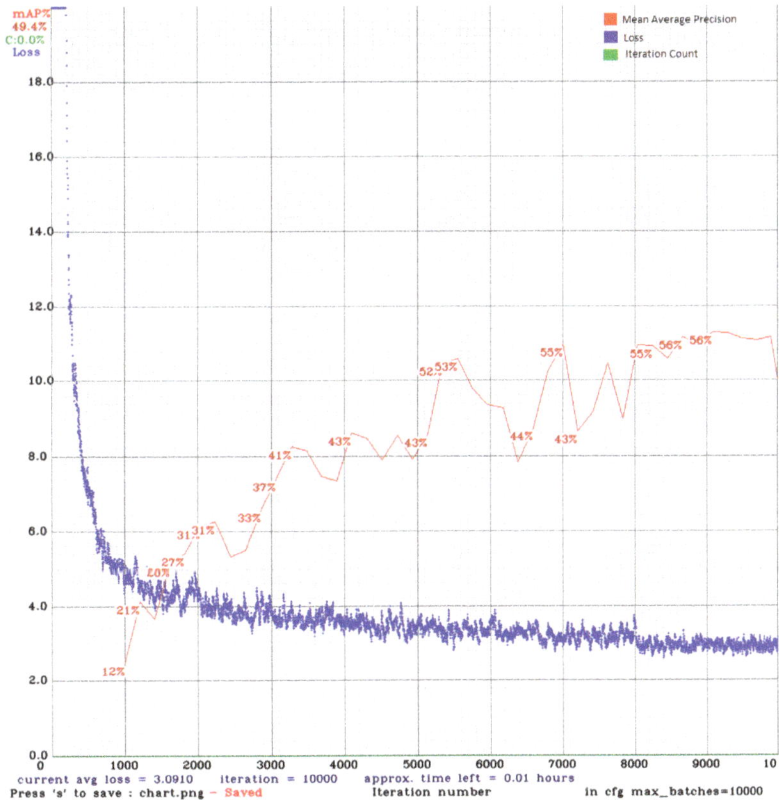

Figure 2. Tiny-YOLOv4 model mAP @ 0.5 (red), loss (blue), iterations (green).

The real-time detection results of the Tiny-YOLOv4 model are shown in Figure 3. The model predicts the potato weeds correctly and efficiently with a high confidence score.

Figure 3. Real-time detection results of the Tiny-YOLOv4 model.

4. Conclusions

DL-based weed detection and classification techniques play a vital role in the domain of agriculture. In this work, a Tiny-YOLOv4 model was used for the detection of potato weeds in a real-time environment. The adopted model gives 49.4% testing accuracy on a very limited dataset. The best-trained model was used for weed detection in potato crops. There is still room for improvement; therefore, we will extend this system for the site-specific spraying technology and will use the most accurate detection algorithm with higher detection accuracy.

Author Contributions: Conceptualization, F.K. and N.Z.; methodology, F.K.; validation, M.A., F.K. and N.Z.; formal analysis, F.K.; investigation, S.S.; resources, S.S. and M.N.T.; data acquisition, F.K. and Z.H.; writing—original draft preparation, F.K. and F.K.; writing—review and editing, F.K. and M.A.; visualization, F.K. and M.A.; supervision, N.Z.; project administration, M.N.T.; funding acquisition, F.K. All authors have read and agreed to the published version of the manuscript.

Funding: The study is a part of PSDP-funded project No. PSDP-332 "Data-Driven Smart Decision Platform" to increase productivity.

Institutional Review Board Statement: Not Applicable.

Informed Consent Statement: Not Applicable.

Data Availability Statement: Not Applicable.

Acknowledgments: The authors would like to thank the DDSDP project.

Conflicts of Interest: The authors declare no conflict of interest.

References

1. Gu, J.; Wang, Z.; Kuen, J.; Ma, L.; Shahroudy, A.; Shuai, B.; Liu, T.; Wang, X.; Wang, G.; Cai, J.; et al. Recent advances in convolutional neural networks. *Pattern Recognit.* **2018**, *77*, 354–377. [CrossRef]
2. Subeesh, A.; Bhole, S.; Singh, K.; Chandel, N.S.; Rajwade, Y.A.; Rao, K.V.R. Deep convolutional neural network models for weed detection in polyhouse grown bell peppers. *Artif. Intell. Agric.* **2022**, *6*, 47–54. [CrossRef]
3. Astani, M.; Hasheminejad, M.; Vaghefi, M. A diverse ensemble classifier for tomato disease recognition. *Comput. Electron. Agric.* **2022**, *198*, 107054. [CrossRef]
4. Huang, M.L.; Chuang, T.C.; Liao, Y.C. Application of transfer learning and image augmentation technology for tomato pest identification. *Sustain. Comput. Inform. Syst.* **2022**, *33*, 100646. [CrossRef]

5. Azimi, S.; Kaur, T.; Gandhi, T.K. A deep learning approach to measure stress level in plants due to Nitrogen deficiency. *Measurement* **2021**, *173*, 108650. [CrossRef]
6. Mishra, A.M.; Harnal, S.; Gautam, V.; Tiwari, R.; Upadhyay, S. Weed density estimation in soya bean crop using deep convolutional neural networks in smart agriculture. *J. Plant Dis. Prot.* **2022**, *129*, 593–604. [CrossRef]

environmental sciences proceedings

MDPI

Proceeding Paper

Development of Real Time Seed Depth Control System for Seeders †

Tahir Iqbal [1,2,*], Abdul Mannan Ghaffar [1], Ubaid ur Rehman [1], Yasir Saeed [3], Muhammad Aqib [3], Fahad Iqbal [3] and Shoaib Rashid Saleem [1,2]

1 Faculty of Agricultural Engineering and Technology, PMAS-Arid Agriculture University, Rawalpindi 46000, Pakistan
2 Data Driven Smart Decision Platform, PMAS-Arid Agriculture University, Rawalpindi 46000, Pakistan
3 National Center for Industrial Biotechnology, PMAS-Arid Agriculture University, Rawalpindi 46000, Pakistan
* Correspondence: tahir.iqbal@uaar.edu.pk; Tel.: +92-333-3975194

† Presented at 1st the International Precision Agriculture Pakistan Conference 2022 (PAPC 2022)—Change the Culture of Agriculture, Rawalpindi, Pakistan, 22–24 September 2022.

Abstract: Proper and uniform seed depth within the optimum range increases crop germination rate, which directly affects yield. Seed depth mainly depends on seed placement techniques. In conventional seed drill and planters, variations and fluctuations in seeding depth were recorded. This issue can be resolved by adaption of intelligent crop sowing machinery. A real-time seed depth control system was designed and developed to monitor and maintain the required seed depth. The developed system was installed on a tractor-mounted conventional seed drill. The materials used were ultrasonic sensors, a micro-controller, an RTC module, buzzers, an LCD, an SD card module, a power battery, and switch buttons. The system read height difference as a reference value. The seeding depth data was stored in the SD card. The results showed by introducing real time depth control system in existing seeders, it is possible to achieve desired plant population and homogeneous germination. The average cumulative germination rate (GR) of wheat with the seed depth control system was 126 plants/m^2, and with a conventional seed drill (without SDCS) 117 plants/m^2. The yield of wheat with the seed depth control system was 1225 kg/acre, and with the conventional seed drill (without SDCS) 1124 kg/acre in an arid region. Result shows that the crop yield could be enhanced by proper seed placement at the required seed depth by using the seed depth control system. Further field trials on different soils, regions and environmental conditions could improve the results.

Keywords: digital agriculture; germination rate; monitoring; seeding depth; yield

Citation: Iqbal, T.; Ghaffar, A.M.; ur Rehman, U.; Saeed, Y.; Aqib, M.; Iqbal, F.; Saleem, S.R. Development of Real Time Seed Depth Control System for Seeders. *Environ. Sci. Proc.* **2022**, *23*, 7. https://doi.org/10.3390/environsciproc2022023007

Academic Editor: Muhammad Naveed Anjum

Published: 28 November 2022

1. Introduction

Food security is a major concern in third-world countries. Considering these countries' continuously expanding populations, expanding industry, and expanding economy, agriculture must develop. The most widely produced cereal crop, and a key source of food in Pakistan, is wheat. It has a yield average of 2600 kg/ha and is grown on a little over 9 million hectares. However, progressive farmers can achieve up to 5000 kg/ha [1]. The low yield is the consequence of several factors. It is primarily caused by low-quality seed, broadcast sowing techniques, late sowing, and insufficient irrigation water. The recommended seed sowing depth for wheat is 3 to 6 cm.

Wheat is sown by five methods: broadcasting, sown behind a local plough, drilling, dibbling and the zero tillage method. These methods result in improper and non-uniform seed distribution. Although drilling can resolve this problem to some extent, it is not fully appropriate. Seed drills are useful for sowing crops such as wheat, barley, sorghum, and maize where there is no need to maintain a certain distance between plants in the same row, while a seed planter is helpful for groundnut [2].

In all the sowing methods for wheat, seeding depth has a great impact on plant height, seed germination rate and yield [3]. If seed depth is in the optimum range, height and seed germination rate should be uniform and homogeneous [4]. This uniform germination rate directly affects the yields of the crops. Therefore, we developed a system that attaches to a conventional seed drill to monitor and maintain the seeding depth and affects plant height, germination rate and yield.

2. Material and Methods

The system was designed and developed at a farm machinery workshop, (33.6492° N, 73.0815° E) FAE&T and tested at Koont Research Farm, (33.1166° N, 73.0111° E) Chakwal. The developed real-time monitoring system program was integrated with all components, i.e., RTC (Real Time clock) module, SD card module, buzzers, LCD, and switch buttons that were electronically connected with a microcontroller through electronic circuits. Ultrasonic sensors were also integrated for measuring depth [5]. After the integration of the microcontroller, the whole system was installed on a tractor-propelled conventional seed drill. Metal plates were used for reference points for the sensors. Tines of a conventional seed drill were placed on the soil surface related to the ON switch of the system. The LCD showed sensor reading as reference values, with respect to the metal plate. Height difference readings were compared, one measured manually from the sensor to metal plates and the other with the sensor. If both values were same, sowing was started, this error was corrected by adjusting the height difference. The system read the height difference as a reference value. The seed depth data were stored on an SD card, inserted into the system. The seed depth control system was tested, and various parameters were monitor and measured i.e., seeding depth with and without the control system on a conventional seed drill, germination rates, and yields.

3. Results and Discussion

3.1. Seeding Depth

Figure 1 shows seeding depth fluctuations in rows during conventional seed drilling without the seed depth control system. The red line represents the maximum permissible seeding depth, while the yellow line shows the minimum permissible seeding depth. As per data collected from five different rows, each row shows a different trend of seeding depth. Seeding depth in these fields were not uniform or homogeneous. In row 1, 2, many seeds were planted too deeply (above 6 cm), and in row 3 and row 5 most of the seed were planted too shallowly (below 3 cm). Overall, 64% of seeds were planted with a permissible seeding depth, while 24% were planted either too deeply (19%) or too shallowly (17%).

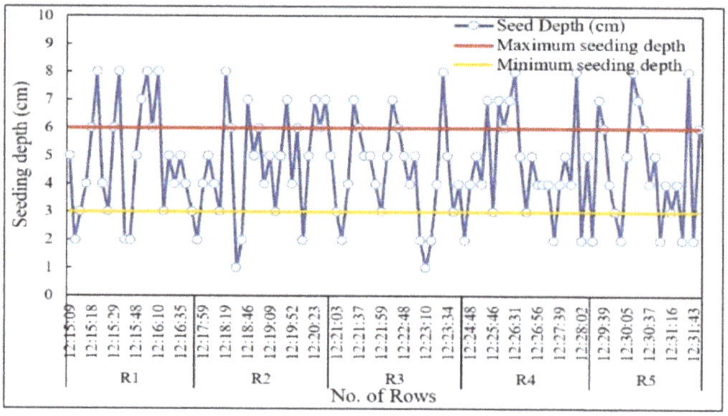

Figure 1. Seeding depth range without SDCS during wheat sowing 2021.

Seeding depth directly affects germination rate, as well as yield. Figure 2 shows the trend of seeding depth in different rows using the seed depth control system. Most of the seeds were planted within the permissible range. Using this system, 94% of seeds were planted within the permissible seeding depth while 6% were planted either too deeply (1.7%) or too shallowly (4.3%).

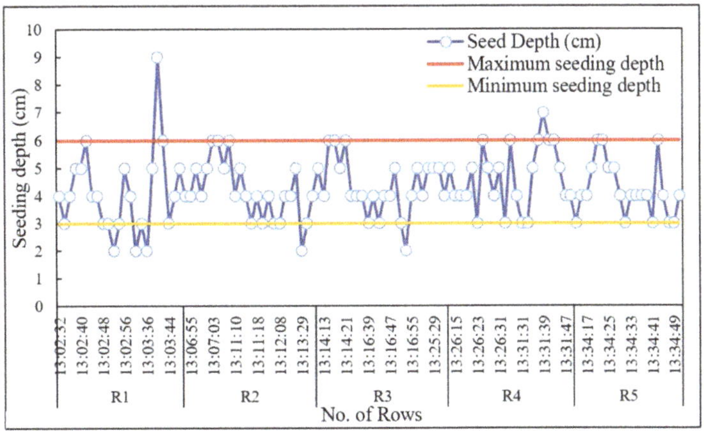

Figure 2. Seeding depth range with SDCS during wheat sowing 2021.

At the maximum value of seeding depth recorded (9 cm), the seed was too deep in the soil, needing more energy to germinate. A minimum value of 2 cm was found at five different positions. Shallow seeds are exposed more to the sunlight, and do not get enough moisture to germinate. These shallow seeds are attacked by birds and rats because they are not placed at the desired depth. By introducing SDCS, most seeds were sown at the proper depth. Figure 1 shows that except R1, the remaining rows (R2, R3, and R4) show homogeneous seeding depth within the required range. These results are in line with the literature [6], and the variation in the seeding depth was quite similar [7].

3.2. *Germination Rate*

The germination rate data in both plots were analyzed. The data were collected on the 7th, 10th, 13th, 16th, 19th, 22nd and 25th day after sowing in both fields. The average accumulative germination rate of wheat without and with the SDCS of 25 samples were 57.17% and 63.04%, respectively The germination rates using SDCS system were significantly improved. Better results could be attained by using hybrid or high-quality seed.

3.3. *YIELD*

The wheat yield was 1225 kg/acre with SDCS, and 1124 kg/acre with the conventional seed drill. Wheat production with SDCS in one hectare was 3028 kg, and with the conventional seed drill was 2778 kg/ha. In last year it was observed 2339 kg/ha. There was a significantly 8% increase in yield achieved by controlling the seeding depth at uniform level.

4. Conclusions

In Pakistan, proper seed depth and uniform distribution of seeds are major problems. Although conventional drills have been introduced to solve these problems, optimum seed depth cannot be maintained. Our seed depth control system (SDCS) maintains seeding depth at the desired range. To validate this system a study was conducted at the University Research Farm to sow wheat seeds with and without SDCS. The average cumulative germination rate of wheat with SDCS was 126 plants/m^2, and with the conventional drill was 117 plants/m^2. The wheat yield with the SDCS drill was 1225 kg/acre and with the conventional drill was 1124 kg/acre. The crop yield increased 8% by placement of seeds at the recommended depth. In Pakistan an estimated 22 million acres of wheat are sowed per year. This represents 40% of the country's total cultivated land during winter. By adopting this system and optimizing seeding depth, around 2.2 million tons more wheat can be produced, which would directly reduce food shortage and minimize the import of wheat.

Author Contributions: Conceptualization, T.I. and U.u.R.; methodology, T.I. and M.A.; software, Y.S., M.A. and F.I.; validation, A.M.G., S.R.S. and Y.S.; formal analysis, T.I.; investigation, A.M.G.; writing—original draft preparation, T.I. and A.M.G.; writing—review and editing, T.I., S.R.S. and M.A. All authors have read and agreed to the published version of the manuscript.

Funding: This research was funded by National Center for Industrial Biotechnology, and Data Driven Smart Decision Platform, PMAS-Arid Agriculture University, Rawalpindi, Pakistan.

Institutional Review Board Statement: Not applicable.

Informed Consent Statement: Not applicable.

Data Availability Statement: Not applicable.

Acknowledgments: The authors are very thankful to the Faculty of Agricultural Engineering and Technology and Univ. Research Farm Administration of PMAS Arid Agriculture University Rawalpindi, Pakistan for providing the platform for this study.

Conflicts of Interest: The authors declare no conflict of interest.

References

1. Finance Division Government of Pakistan. Pakistan Economic Survey 2019-20. In *Economic Survey. Islamabad: Economic Advisor's Wing, Ministry of Finance.*; 2020. Available online: https://www.finance.gov.pk/survey/chapter_20/PES_2019_20.pdf (accessed on 13 March 2022).
2. Sarker, T.R.; Alam, M.; Ali, M.R.; Saha, C.K.; Haque, M.A. Design and development of a multi-crop manual seed drill. *CIGR J.* **2019**, *21*, 51–60.
3. Sedara, A.; Ibrahim, Y.; Manuwa, S.; Sedara, O. Development of a Low-Cost Seed Drill for Sowing Amaranth Seed Development of a Low-Cost Seed Drill for Sowing Amaranth Seed. *World Acad. J. Eng. Sci.* **2020**, *7*, 39–45.
4. Nielsen, S.K.; Munkholm, L.J.; Lamandé, M.; Nørremark, M.; Edwards, G.T.C.; Green, O. Seed drill depth control system for precision seeding. *Comput. Electron. Agric.* **2017**, *144*, 174–180. [CrossRef]
5. Dvorak, J.S.; Stone, M.L.; Self, K.P. Object detection for agricultural and construction environments using an ultrasonic sensor. *J. Agric. Saf. Health.* **2016**, *22*, 107–119. [CrossRef] [PubMed]
6. Suomi, P.; Oksanen, T. Automatic working depth control for seed drill using ISO 11783 remote control messages. *Comput. Electron. Agric.* **2015**, *116*, 30–35. [CrossRef]
7. Nielsen, S.K.; Munkholm, L.J.; Lamandé, M.; Nørremark, M.; Skou-Nielsen, N.; Edwards, G.T.C.; Green, O. Seed drill instrumentation for spatial coulter depth measurements. *Comput. Electron. Agric.* **2017**, *141*, 207–214. [CrossRef]

environmental sciences proceedings

MDPI

Proceeding Paper

Modeling Irrigation Water Requirement of Mixed Crop with Coupled Smart Irrigation System and System Dynamic Model †

Fiaz Hussain [1] and Ray-Shyan Wu [2,*]

1 Department of Agricultural Engineering, PMAS-Arid Agriculture University, Rawalpindi 46300, Pakistan
2 Department of Civil Engineering, National Central University, Chung-Li, Taoyuan 32001, Taiwan
* Correspondence: raywu@ncu.edu.tw
† Presented at the 1st International Precision Agriculture Pakistan Conference 2022 (PAPC 2022)—Change the Culture of Agriculture, Rawalpindi, Pakistan, 22–24 September 2022.

Abstract: Water is a key component in the two biggest economic drivers of Taiwan, i.e., the semiconductor and agricultural industries. Agricultural water accounts for 70% of the total water usage of the nation. During drought situations, the allocation and utilization of agricultural water usage become an important issue where farmers pump groundwater to supply the irrigation deficit from surface water, which ultimately impacts regional groundwater levels. Thus, there is a need to find a way to address its field water consumption during droughts; one way is a smart irrigation water management system. In this study, a smart irrigation water management model coupled with a system dynamic model (VENSIM) was developed for mixed crops in Central Taiwan by reducing 50% of the planned irrigation. Results can be applied as a solution to water shortage during droughts with alternate frequent adjustment of water gates to ensure water supply to tail end users.

Keywords: VENSIM model; precision irrigation; droughts; irrigation plan; paddy field

Citation: Hussain, F.; Wu, R.-S. Modeling Irrigation Water Requirement of Mixed Crop with Coupled Smart Irrigation System and System Dynamic Model. *Environ. Sci. Proc.* **2022**, 23, 8. https://doi.org/10.3390/environsciproc2022023008

Academic Editor: Qamar Zaman

Published: 28 November 2022

1. Introduction

Taiwan has been facing water stress since late 2020 due to extended droughts and changing precipitation patterns under climate change, which is a wake-up call under current scenarios to devise a water rationing policy for agriculture, semiconductor, other industry, and domestic purposes. During prolonged droughts water shortages continue to challenge authorities due to Taiwan's uneven rainfall distribution, dense population, storage capacity, and geographical characteristics. Furthermore, global climate changes continue to worsen the current shortage situation and present unprecedented challenges to Taiwan's water system [1]. The persistent lack of rainfall leads to a stoppage of allocated agricultural water and transferred to industry which created unrest conditions for farmers. In 2021, 74,000 ha area of first season rice crops was deprived of irrigation water which accounts for 24% of total planted area [2]. In 2015, 43 thousand acres of paddy field did not receive water for irrigation and the country's reservoir levels dipped below 50%. Taiwan is experiencing worst drought in 56 years and no typhoon passed to refill its reservoirs and level dropped down to 30% in 2021 [3].

There must be an alternative way to manage water supply equitability for all sectors because the severity of droughts is increasing and cutting down water demand for farming and chipmaking is not a viable solution [4]. The farmers should adopt new irrigation methods such as precision irrigation and smart irrigation gates. There should be the adaptation of a sensor-based soil moisture measurement smart system for making agricultural water usage more efficient. This study used system dynamic program VENSIM [5], to develop a smart irrigation system for water management with a 30% and 50% reduction of allocation in the study region of Central Taiwan. The study objective is to develop a smart irrigation water management model coupled with VENSIM simulation tool during drought in Central Taiwan.

2. Study Area Description

The site is located in Chang-Hua County having an area of 215 ha which is divided into five blocks (Figure 1). The mixed cropping system consists of paddy fields (70%) and upland crops (cabbage, 30%). There are five irrigation channels and six field water level monitoring stations.

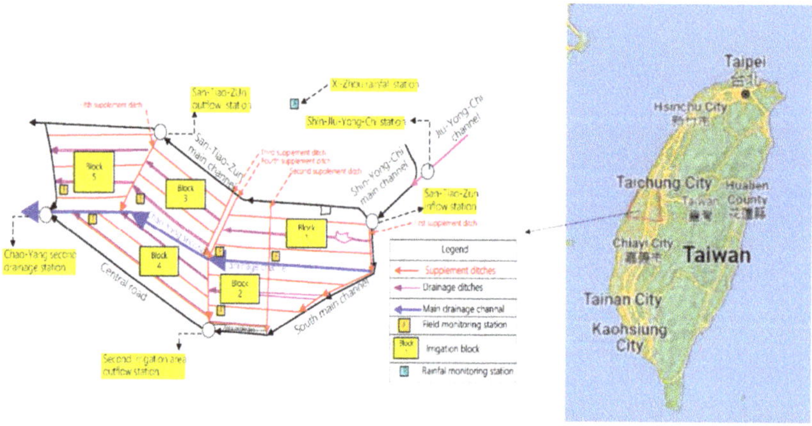

Figure 1. The layout of the experimental site.

3. Material and Methods

The water balance method is conceptualized as shown in Figure 2. The VENSIM model was formulated using mathematical governing equations. The sensors detect water levels and transfer the information to the data center at every 10 min (Figure 3). The crop water requirement model obtains data from the data center and calculates field overflow, infiltration, and evapotranspiration to estimate irrigation demand, which is then transported to the data center within a 2 h cycle.

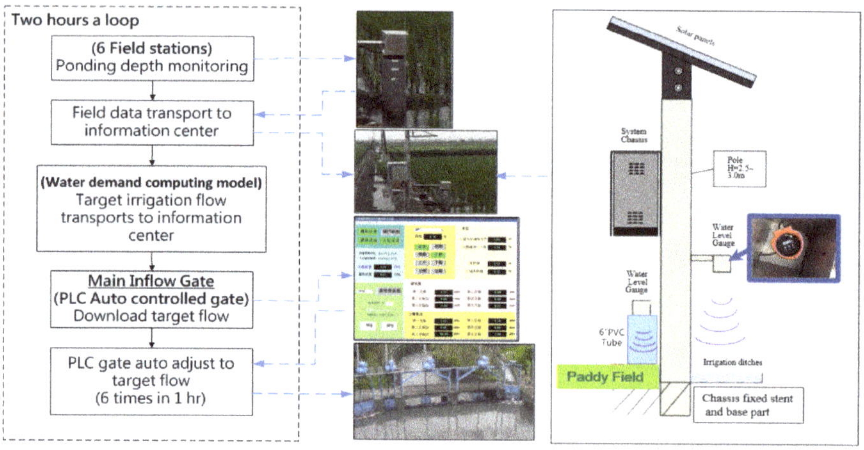

Figure 2. Smart irrigation system equipped with water level sensors and control unit [1].

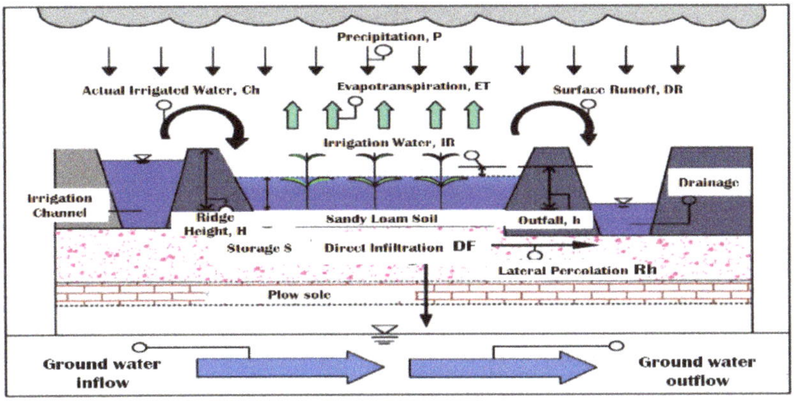

Figure 3. The conceptual model used to develop water balance approach [1].

4. Results

The model was verified with R2 = 0.83 and simulation results are shown in Figure 4. During 50% reduction, the fourth block cannot have sufficient water for the target depth during survival period; however, it obtains the target depth on the 34th day tillering stage. The depth of the fifth block reached below the saturated soil moisture on the 6th day from transplanting due to the shortage of water to the 31st day. On the 21st day, the field storage becomes lower than the field capacity (FC), and the vertical percolation stops. On the 29th day, the field storage becomes much lower and reaches the wilting point (WP), stopping evapotranspiration.

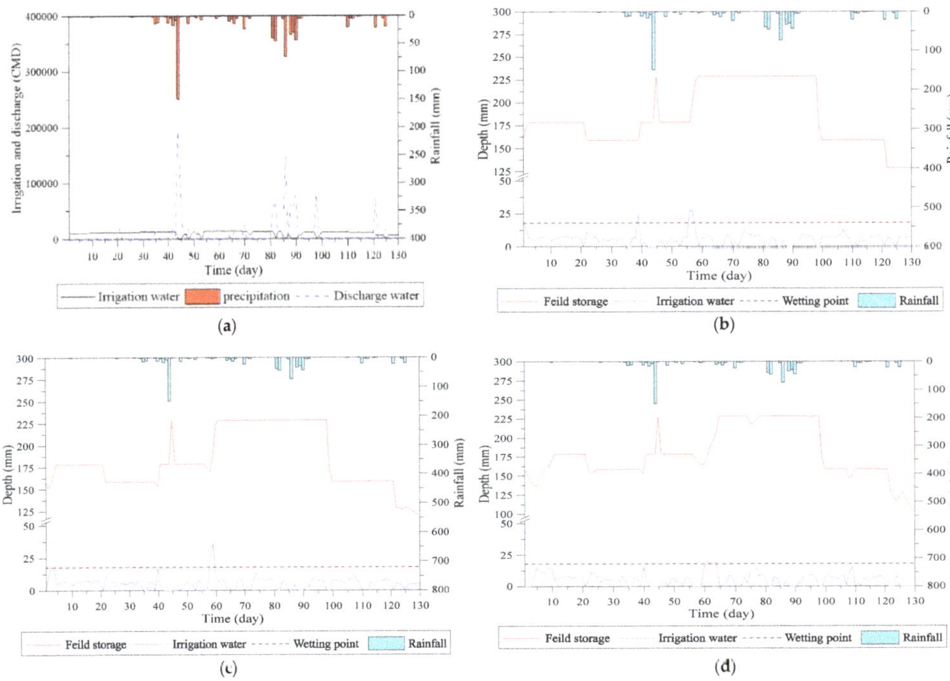

Figure 4. *Cont.*

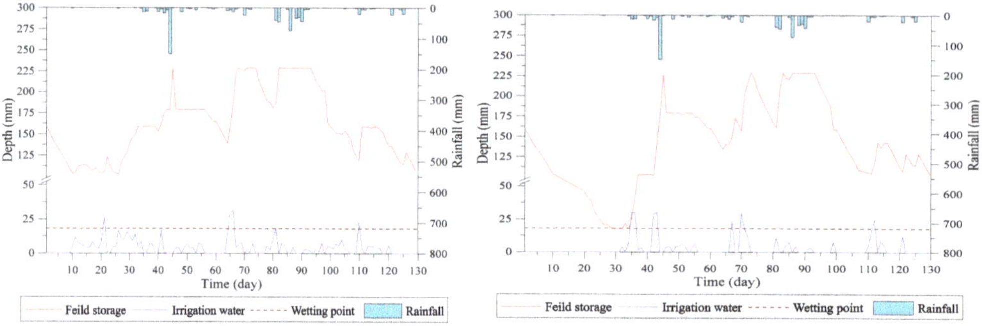

(e) (f)

Figure 4. The simulation of a 50% drop of planned irrigation (**a**) and simulation of blocks 1 to 5 (**b–f**).

5. Conclusions

The water balance model was used to simulate an experimental site's water demand and supply to analyze water saving during drought. Stricter with drought, the scenario of a 50% reduction in irrigation water can be applied as a solution to water shortage. As a result of the study, before the 21st day, block 5 should be irrigated to avoid its field storage being lower than FC. It is suggested that when applying a 50% discount policy of provided irrigation water, every block should be irrigated in turns. When the irrigation turn belongs to the downstream area, the water gates located upstream should be adjusted recurrently to guarantee the downstream blocks can receive the allocated water rather than making irrigation water overflow in the front blocks. This model can be applied as a solution to water shortage during droughts for agricultural water management smartly by farmers.

Author Contributions: Conceptualization and methodology, F.H. and R.-S.W.; writing—original draft preparation, F.H.; writing—review and editing, R.-S.W. All authors have read and agreed to the published version of the manuscript.

Funding: This research received no external funding.

Institutional Review Board Statement: Not applicable.

Informed Consent Statement: Not applicable.

Data Availability Statement: Not applicable.

Conflicts of Interest: The authors declare no conflict of interest.

References

1. Wu, R.-S.; Liu, J.-S.; Chang, S.-Y.; Hussain, F. Modeling of Mixed Crop Field Water Demand and a Smart Irrigation System. *Water* **2017**, *9*, 885. [CrossRef]
2. Lin, O.; Syngle, E. *Taiwan Drought Results in Rice Area Reduction and Crop Loss*; TW2021-0037; Voluntary—Public Distribution; USDA-Foreign Agricultural Service: Taipei, Taiwan, 2021.
3. Commonwealth Magazine. Available online: https://web.cw.com.tw/drought-2021-en/index.html (accessed on 27 September 2022).
4. Aljazeera. Available online: https://www.aljazeera.com/news/2021/8/20/taiwan-water-woes (accessed on 27 September 2022).
5. Ventana Systems, Inc. Vensim User's Guide Version 5 [EB/OL]. USA. Available online: http://vensim.com/ (accessed on 27 September 2022).

environmental sciences proceedings

MDPI

Proceeding Paper

Rainwater Harvesting: A Sustainable Water Management Option for Irrigation of Public Parks †

Muhammad Sohail Waqas [1,*], Muhammad Jehanzeb Masud Cheema [2,3], Saddam Hussain [4,5], Muhammad Usman Khalid [6] and Muhammad Saadullah Khan [3]

1 Soil Conservation Group, Agriculture Department (Field Wing), Government of the Punjab, Rawalpindi 46000, Pakistan
2 Faculty of Agricultural Engineering and Technology, PMAS-Arid Agriculture University, Rawalpindi 46000, Pakistan
3 National Center for Industrial Biotechnology, PMAS-Arid Agriculture University, Rawalpindi 46000, Pakistan
4 Department of Irrigation and Drainage, University of Agriculture Faisalabad, Faisalabad 38000, Pakistan
5 Department of Agricultural and Biological Engineering, University of California, Davis (UC Davis), Davis, CA 95616, USA
6 Program Management Unit, Agriculture Department, Government of the Punjab, Rawalpindi 46000, Pakistan
* Correspondence: sohailwaqasrana@gmail.com

† Presented at the 1st International Precision Agriculture Pakistan Conference 2022 (PAPC 2022)—Change the Culture of Agriculture, Rawalpindi, Pakistan, 22–24 September 2022.

Abstract: Water is imperative for life and plays a vital role in sustaining multiple environmental services. Currently, water resources are under stress, and rainwater harvesting (RWH) is one of the solutions available to address water shortages. In this study, the potential of RWH for irrigation and recreational activities in a public park (i.e., Fatima Jinnah Park-Islamabad) is discussed. The soil conservation service-curve number (SCS-CN) method was used to estimate the runoff. Results revealed that annually 1.80 million cubic meters (MCM) runoff is generated at this park. If this runoff volume is accommodated and managed effectively, it can fulfill the requirements of irrigation and other water-related activities. Hence, the adoption of RWH technology is vital for managing water; therefore, this approach should be used to support any policy changes that lead to widespread use of RWH.

Keywords: runoff; water conservation; groundwater depletion; curve number; urban stormwater management

Citation: Waqas, M.S.; Cheema, M.J.M.; Hussain, S.; Khalid, M.U.; Khan, M.S. Rainwater Harvesting: A Sustainable Water Management Option for Irrigation of Public Parks. *Environ. Sci. Proc.* **2022**, 23, 9. https://doi.org/10.3390/environsciproc2022023009

Academic Editor: Shoaib Rashid Saleem

Published: 28 November 2022

Publisher's Note: MDPI stays neutral with regard to jurisdictional claims in published maps and institutional affiliations.

1. Introduction

Water, being the most valued resource on earth [1] is considered to be an integral development index of a society [2]. Unfortunately, ongoing depletion of this valuable resource has caused worldwide water scarcity. The major user of the water resources related to the production of vital food crops is agriculture, while inaccessibility to water is limiting agricultural production [3–5]. Similarly, the level of groundwater decreases as a result of increased groundwater exploitation for domestic and commercial irrigation purposes. Therefore, the judicious utilization of water resources is the need of the hour [6]. This can only be achieved by using available water supplies efficiently and adopting conservation practices. Therefore, rainwater harvesting (RWH) is an answer to water scarcity at all levels. The RWH technology is a relatively low-cost option [7], and highly decentralized which empowers individuals and communities to manage their water [8,9]. Mostly, recreational activities (i.e., fountains, etc.) and irrigation of public parks are carried out by pumping water from underground aquifers; because of this, groundwater reserves are being depleted rapidly. In this context, the RWH technique is the prominent adaptive strategy for managing water efficiently. The current study was designed to evaluate the importance of RWH system for irrigation and other water-related activities at public parks.

2. Materials and Methods

2.1. Study Area

To assess the potential of RWH, Fatima Jinnah Park (F9-Park) was chosen. This park is located in Islamabad Capital Territory (ICT), the location coordinates of the site are 33.702433°N, 73.023105°E (Figure 1).

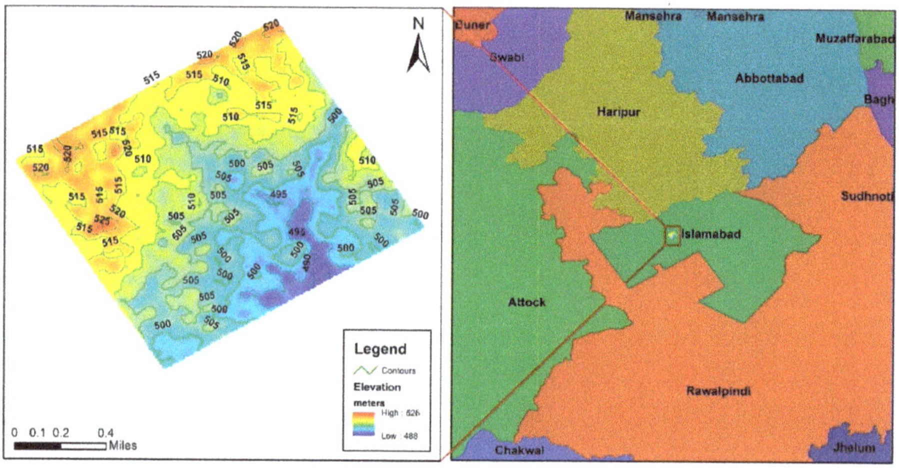

Figure 1. Location Map of Fatima Jinnah Park (F9-Park) ICT, showing the elevation of the study area along with contours at 2 m interval.

2.2. Runoff Calculation

Runoff depth is used to assess the potential water supply during runoff. The soil conservation service-curve number (SCS-CN) method was used to estimate the runoff, which can be expressed as follows:

$$Q = \frac{(P - I_a)^2}{(P - I_a) + S} \tag{1}$$

where Q is runoff depth (mm), P is precipitation (mm), I_a is an initial abstraction (mm), and S is potential maximum retention (mm). Whereas, according to [10] $I_a = 0.2\ S$. Hence, the above equation (Equation (1)) can be re-write as follows:

$$Q = \frac{(P - 0.2S)^2}{(P + 0.8S)} \tag{2}$$

Potential maximum retention (S) can be determined using the following relation:

$$S = \frac{25400}{CN} - 254 \tag{3}$$

CN varies from 0 to 100 and represents the runoff response to a given rain. High CNs indicate that a large proportion of the rainfall will become surface runoff [2]. The curve number was obtained (Table 1), after selecting a land-use type and a corresponding hydrologic soil group according to (USDA-SCS, 1985).

Table 1. Runoff Calculation at Fatima Jinnah Park (F9-Park) ICT.

Month	Normal Rainfall *	Curve Number	Max. Potential Retention (S)	Runoff Depth (Q)		Catchment Area		Runoff Volume
	mm	-	mm	mm	m	ha	m^2	m^3
January	59.00	67.50	122.30	7.61	0.008	300	3,000,000	22,821
February	89.00	67.50	122.30	22.29	0.022	300	3,000,000	66,885
March	87.70	67.50	122.30	21.56	0.022	300	3,000,000	64,667
April	59.60	67.50	122.30	7.84	0.008	300	3,000,000	23,531
May	38.20	67.50	122.30	1.39	0.001	300	3,000,000	4164
June	78.20	67.50	122.30	16.41	0.016	300	3,000,000	49,218
July	368.60	67.50	122.30	253.91	0.254	300	3,000,000	761,729
August	334.70	67.50	122.30	222.52	0.223	300	3,000,000	667,568
September	123.30	67.50	122.30	44.18	0.044	300	3,000,000	132,535
October	32.70	67.50	122.30	0.52	0.001	300	3,000,000	1561
November	11.90	67.50	122.30	1.44	0.001	300	3,000,000	4312
December	40.40	67.50	122.30	1.84	0.002	300	3,000,000	5515
Total:	1323.30			601.50	0.602			1,804,505

* Normal monthly rainfall assessed from Pakistan Meteorological Department (PMD).

3. Results and Discussion

The total area of the park is around 300 hectares, with 526 m as the maximum elevation and 488 m as the minimum elevation (Figure 1). The summation of normal monthly rainfall of the study area is 1323.30 mm, while maximum rainfall amounted to 368.60 mm observed during the month of July (Table 1). Moreover, an average CN was chosen (i.e., 67.50) given by (USDA-SCS, 1985), and the maximum potential retention of the study area is calculated as 122.30 mm.

The maximum runoff generated during the monsoon period (i.e., July–September), while the annual runoff volume of the park is 1,804,505 m^3. Moreover, the possible location for harvesting rainwater is shown in Figure 2. Many other studies from different regions also reported the importance of using RWH for municipal purposes i.e., park irrigation [11], commercial sites [12], and to reduce on ground water shortages [13,14].

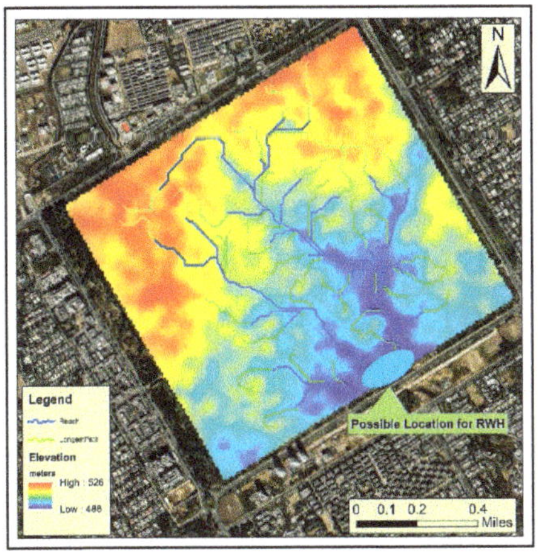

Figure 2. Showing possible location for RWH, reaches, longest flow paths, and elevations.

4. Conclusions

RWH is the collection of rainwater during the rainy season in order to use it later when there is a shortage of water. The present study found that a significant volume of runoff can be harvested, stored, and reused for irrigation of parks. Hence, RWH is a holistic technique for enhancing water resources, and is indispensable in terms of groundwater sustainability and protecting the natural ecosystem.

Author Contributions: Conceptualization, M.S.W., M.J.M.C. and S.H.; methodology, M.S.W., M.J.M.C. and S.H.; software, M.S.W.; validation, M.S.W., M.J.M.C., S.H., and M.U.K.; formal analysis, M.S.W. and M.J.M.C.; investigation, M.S.W. and S.H.; resources, M.S.W., S.H. and M.S.K.; data curation, M.S.W., S.H., M.U.K. and M.S.K.; writing—original draft preparation, M.S.W.; writing—review and editing, M.S.W., M.J.M.C. and S.H.; visualization, M.S.W.; supervision, M.J.M.C. All authors have read and agreed to the published version of the manuscript.

Funding: This research received no external funding.

Institutional Review Board Statement: Not applicable.

Informed Consent Statement: Not applicable.

Data Availability Statement: Not applicable.

Conflicts of Interest: The authors declare no conflict of interest.

References

1. Hossain, M.Z. Water: The most precious resource of our life. *Glob. J. Adv. Res.* **2015**, *2*, 1436–1445.
2. Khoso, S.; Wagan, F.H.; Tunio, A.H.; Ansari, A.A. An overview on emerging water scarcity in Pakistan, its causes, impacts and remedial measures. *J. Appl. Eng. Sci.* **2015**, *13*, 35–44. [CrossRef]
3. Waqas, M.S.; Cheema, M.J.M.; Hussain, S.; Ullah, M.K.; Iqbal, M.M. Delayed irrigation: An approach to enhance crop water productivity and to investigate its effects on potato yield and growth parameters. *Agric. Water Manag.* **2021**, *245*, 106576. [CrossRef]
4. Iqbal, M.M.; Hussain, S.; Cheema, M.J.M.; Lyul, J. Seasonal effect of agricultural pollutants on coastline environment: A case study of the southern estuarine water ecosystem of the boseong county Korea. *Pak. J. Agri. Sci.* **2022**, *59*, 117–124.
5. Dilawar, A.; Chen, B.; Ashraf, A.; Alphonse, K.; Hussain, Y.; Ali, S.; Jinghong, J.; Shafeeque, M.; Boyang, S.; Sun, X.; et al. Development of a GIS based hazard, exposure, and vulnerability analyzing method for monitoring drought risk at Karachi, Pakistan. *Geomat. Nat. Hazards Risk* **2022**, *13*, 1700–1720. [CrossRef]
6. Waqas, M.; Cheema, M.J.M.; Hussain, S.; Daccache, A.; Ullah, M.K.; Iqbal, M.M.; Hussain, H. Impact of advanced and delayed irrigation practices on physico-chemical attributes of potato tubers under bed sowing. *J. Anim. Plant Sci.* **2022**, *32*, 705–720.
7. Aftab, T.B.; Hasnain, S.A.; Iqbal, S.R. Save water and safe water; evaluation of design and storage period on water quality of rainwater harvesting system. *J. Environ. Earth Sci.* **2012**, *2*, 145–160.
8. Ali, A. Water harvesting in dry environments. In *Innovations in Dryland Agriculture*; Springer: Cham, Switzerland, 2016; pp. 49–98.
9. Iqbal, M.M.; Li, L.; Hussain, S.; Lee, J.L.; Mumtaz, F.; Elbeltagi, A.; Waqas, M.S.; Dilawar, A. Analysis of Seasonal Variations in Surface Water Quality over Wet and Dry Regions. *Water* **2022**, *14*, 1058. [CrossRef]
10. Melesse, A.M.; Shih, S. Spatially distributed storm runoff depth estimation using Landsat images and GIS. *Comput. Electron. Agric.* **2002**, *37*, 173–183. [CrossRef]
11. Nachshon, U.; Netzer, L.; Livshitz, Y. Land cover properties and rain water harvesting in urban environments. *Sustain. Cities Soc.* **2016**, *27*, 398–406. [CrossRef]
12. Mohammed, T.A.; Noor, M.; Ghazali, A.H. Study on potential uses of rainwater harvesting in urban areas. In Proceedings of the Colloquium on Rainwater Utilization, Putrajaya, Malaysia, 19–20 April 2007; pp. 1–15.
13. Krishna, R.; Mishra, J.; Ighalo, J.O. rising demand for rain water harvesting system in the world: A case study of Joda Town, India. *World Sci. News* **2020**, *146*, 47–59.
14. Cheema, M.; Bastiaanssen, W.G. Land use and land cover classification in the irrigated Indus Basin using growth phenology information from satellite data to support water management analysis. *Agric. Water Manag.* **2010**, *97*, 1541–1552. [CrossRef]

environmental sciences proceedings

MDPI

Proceeding Paper

Experimental Study on the Effect of Convective Drying of Potato Slices with Sequentially Reducing Temperature †

Muhammad Usama *, Muhammad Khizar Hayyat, Ammar Ahmed, Muhammad Ali, Zaib Ali and Aashir Iqbal

School of Mechanical and Manufacturing Engineering, National University of Science and Technology (NUST), Islamabad 44000, Pakistan
* Correspondence: usama@bfreeze.com
† Presented at the 1st International Precision Agriculture Pakistan Conference 2022 (PAPC 2022)—Change the Culture of Agriculture, Rawalpindi, Pakistan, 22–24 September 2022.

Abstract: Solar convective drying is a method of dehydrating food that is gaining popularity in developing regions due to its low power consumption and shorter yield times compared to direct sun drying. Exposure of food items to high temperatures towards the end of drying results in color and shape deterioration, negatively affecting the product's market value. To alleviate this problem, we explored the impact of dehydrating potato slices using Convective Drying with reducing temperatures over the drying process. It was found that reducing the temperature in two steps during the drying process preserved 61% of the original color at the cost of a 23.8% increase in drying time, compared to constant temperature drying at 60 °C.

Keywords: convection; solar drying; drying time; potato drying; food systems; scorching; burning; food preservation; food drying

Citation: Usama, M.; Hayyat, M.K.; Ahmed, A.; Ali, M.; Ali, Z.; Iqbal, A. Experimental Study on the Effect of Convective Drying of Potato Slices with Sequentially Reducing Temperature. *Environ. Sci. Proc.* **2022**, 23, 10. https://doi.org/10.3390/environsciproc2022023010

Academic Editor: Tahir Iqbal Saddam Hussain

Published: 19 December 2022

1. Introduction

A total of 30% of the agricultural produce in developing countries is wasted as a post-harvest loss. While freezing food items is a globally adopted solution to prolong the shelf life of food, refrigeration is energy expensive and thereby eco-hazardous. Dehydration or drying is a more sustainable method of preserving fruits and vegetables to make food systems secure for a rapidly increasing global population.

Drying as a method of food preservation is particularly common in developing countries [1]. This is primarily due to its low operational cost. For this reason, the most popular method of dehydrating fruits and vegetables is Open Sun Drying. This method involves leaving food items in direct sunlight. The moisture present in the food is evaporated due to the vapor pressure difference created by solar heating [2]. Due to long drying times, the quality of nutritional and cosmetic quality of the product is inferior compared to more expensive methods [3]. Solar Convective Drying overcomes some of the shortcomings of Open Sun Drying. This method involves the use of solar thermal collectors to conduct drying at higher temperatures. This reduces the drying time as the moisture removal rate is relatively higher [4]. Due to short drying times and higher color retention, Solar Convective Drying is a more feasible method of dehydrating fruits and vegetables. It has been found that convective drying delivers better product quality at high temperatures [5]. Even so, scorching of food surface and poor color retention is a persistent problem with Convective drying [6].

In this work, we tested the effect of reducing the hot air temperature during the drying process to keep the surface temperatures at values that do not damage the food item.

2. Methodology

Potato slices 3 mm ± 0.5 mm thick, weighing 5 g ± 0.1 g, were used for the experiments. A 40 W centrifugal fan was used to blow air for convection. Airflow was measured with a hot wire anemometer calibrated against a manometer and static pressure. The surface temperature was measured with an infra-red thermometer which was calibrated against phase change points of water. Air temperature and Relative Humidity were measured with an XH-M452 module. Color retention was measured by conducting pixel thresholding of photographs of the potato slices. Figure 1 depicts the schematic of the equipment setup and instrumentation. The 1st experiment was conducted at a constant temperature of 40 °C. The 2nd experiment was conducted at a constant temperature of 60 °C. The 3rd experiment was started at 60 °C and when the surface temperature crossed 40 °C, the air temperature was dropped to 50 °C. Following this temperature reduction, when the surface temperature crossed 45 °C, the air temperature was dropped even further down to 40 °C. The measured parameters were the mass and surface temperature of drying slices.

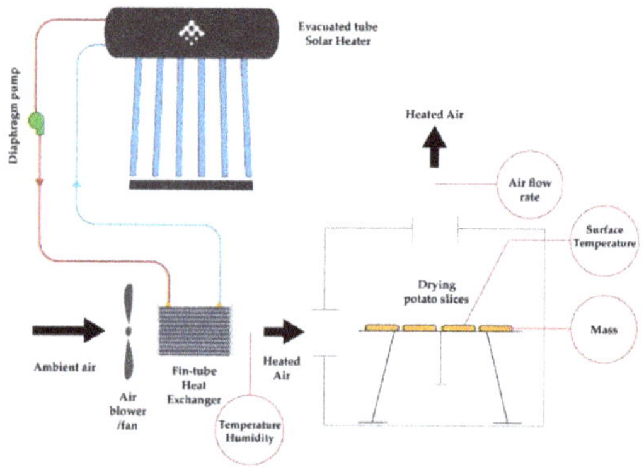

Figure 1. Solar evacuated tube based convection dryer with air flow, temperature, humidity and mass measurement.

3. Theoretical Model

Lewis' model of drying was used to theoretically calculate the rate of dehydration. The drying constant was iterated to fit the data closest to the experimental results to obtain the drying curve as shown in Figure 2.

$$MR(t) = e^{-kt} \tag{1}$$

$$m(t) = m_w(t) + m_d \tag{2}$$

where

$$m_w(t) = (MR)(m_i), \tag{3}$$

$$m_d = (m_i)(1 - MR_i) \tag{4}$$

$$k = drying\ constant,\ MR = Moisture\ Ratio,\ t = time,\ m(t) = mass, \tag{5}$$

$$m_w(t) = water\ mass,\ m_d = dry\ mass, \tag{6}$$

$$m_i = initial\ mass,\ MR_i = initial\ Moisture\ Ratio \tag{7}$$

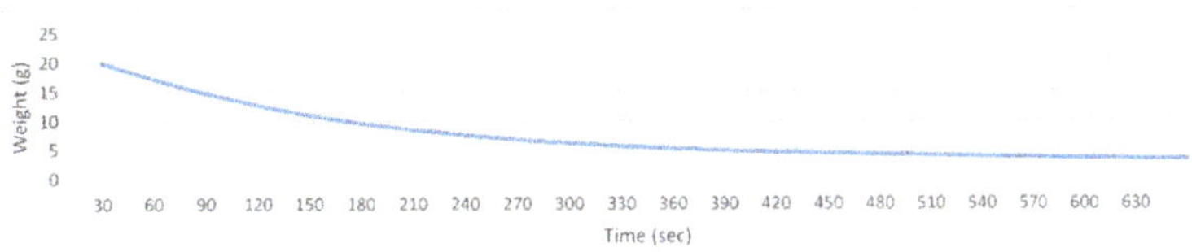

Figure 2. Drying rate predicted by theoretical model for Convective Drying at 60 °C.

4. Results

Table 1 shows that Experiment-2 delivered the fastest drying time and Experiment-3 delivered the highest color retention. Figure 3 shows the plot of mass and surface temperature over the drying period for Experiment-1. Figure 4 shows the plots for Experiment-2 and Figure 5 shows the drying rate and surface temperatures over time for Experiment-3.

Table 1. Experimental parameters and results.

No.	Inlet Temperature (°C)	Air Speed (m/s)	Initial Mass (g)	Final Mass (g)	Drying Time (min)	Color Retention (%)
1	40	1	20	5.1	33	44.9
2	60	1	20	3.9	21	41.2
3	60–50–40	1	20	4.6	26	61.0

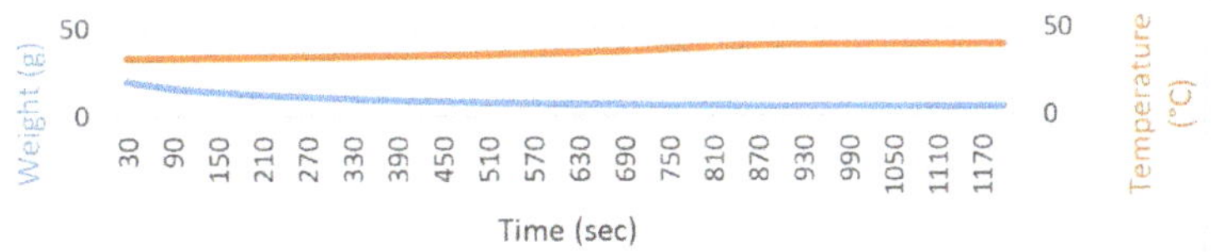

Figure 3. Graphical results for Constant temperature convection at 40 °C.

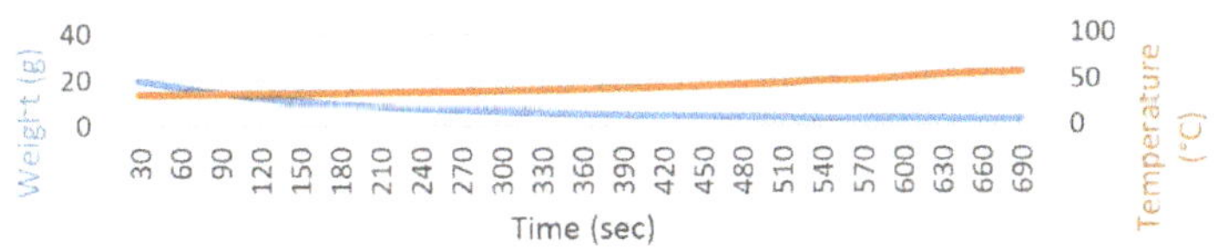

Figure 4. Graphical results for Constant temperature convection at 60 °C.

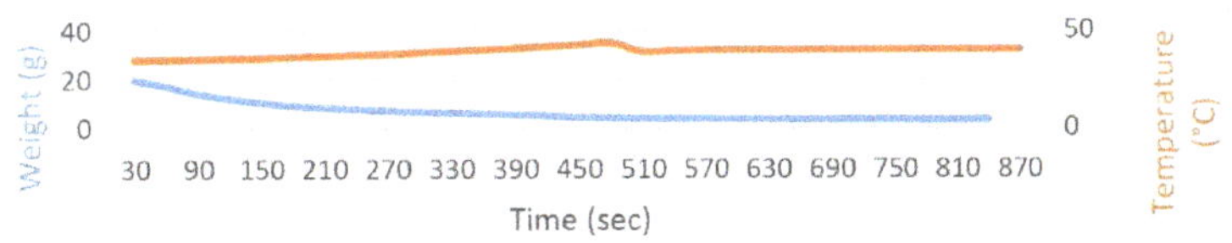

Figure 5. Graphical results for step-reduction in temperature by 10 °C.

5. Conclusions

5.1. Drying Rate

- The drying rate was found to observe an exponential decay for all experiments. This is due to the rate of evaporation at the surface being higher towards the beginning and reducing over time. The moisture that is removed further through the drying process is located further away from the surface.
- For step-reduction of temperature done in Experiment—3, it was found that the slope of the drying rate reduces on each temperature reduction step.

5.2. Surface Temperature

- The surface temperature was found to exponentially converge to the temperature of the hot air. This is because the moisture at the surface acts as a phase-change coolant that depletes over time. Thus, the rate of evaporative cooling at the surface approaches zero as the drying process continues.
- For Experiment—3, the temperature curve was identical to that obtained with constant temperature drying at 60 °C for the first 210 seconds. The slope of the curve reduced to zero as the temperature change became linear after the first step-reduction. The second step-reduction caused a sudden drop in temperature following which it approached and stagnated at 40 °C.

Overall, the surface temperature remained below 45 °C throughout the drying process with a 23.8% increase in drying time over constant temperature drying at 60 °C. Reducing the temperature in two steps resulted in 19.8% higher color retention and thereby improved cosmetic product quality. This makes the product much more suitable to compete with dried snacks made with eco-hazardous and energy-inefficient methods.

Author Contributions: Conceptualization, M.U. and Z.A.; methodology, M.U., M.A. and M.K.H.; manufacturing, M.U., M.A. and A.I.; experimentation, M.U., M.K.H., A.A. and A.I.; data analysis, M.U. and A.I.; drafting, M.U.; supervision, Z.A. All authors have read and agreed to the published version of the manuscript.

Funding: This research received no external funding.

Institutional Review Board Statement: Not applicable.

Informed Consent Statement: Not applicable.

Data Availability Statement: The data points for the experiments conducted can be accessed here: https://www.shorturl.at/shortener.php (accessed on 15 December 2022).

Acknowledgments: The authors would like to thank School of Mechanical and Manufacturing Engineering (SMME) for providing lab, Bfreeze Pvt Ltd. (Islamabad, Pakistan) for providing equipment and the reviewers for providing feedback to refine the manuscript.

Conflicts of Interest: The authors declare no conflict of interest.

References

1. Samykano, M. Progress in one-dimensional nanostructures. *Mater. Charact.* **2021**, *179*, 111373. [CrossRef]
2. Giri, S.K.; Prasad, S. Drying kinetics and rehydration characteristics of microwave-vacuum and convective hot-air dried mushrooms. *J. Food Eng.* **2007**, *78*, 512–521. [CrossRef]
3. Asnaz, M.S.K.; Dolcek, A.O. Comparative performance study of different types of solar dryers towards sustainable agriculture. *Energy Rep.* **2021**, *7*, 6107–6118. [CrossRef]
4. Tulej, W.; Głowacki, S. Modeling of the Drying Process of Apple Pomace. *Appl. Sci.* **2022**, *12*, 1434. [CrossRef]
5. Hasan, M.U.; Malik, A.U.; Ali, S.; Imtiaz, A.; Munir, A.; Amjad, W.; Anwar, R. Modern drying techniques in fruits and vegetables to overcome postharvest losses: A review. *J. Food Process. Preserv.* **2019**, *43*, e14280. [CrossRef]
6. Semai, H.; Bouhdjar, A.; Amari, A. An improved protocol for natural convective drying of pumpkin. *Food Feed. Res.* **2021**, *48*, 29–39. [CrossRef]

environmental sciences proceedings

MDPI

Proceeding Paper

Nutrient Indexing of Different Olive Cultivars under Rainfed Conditions †

Kashif Mehmood Khan [1], Qaiser Hussain [1,*], Muhammad Akmal [1], Muhammad Azam Khan [2], Sarosh Alvi [3], Tayyaba Shakeel [4], Rabia Manzoor [1] and Muhammad Irfan [1]

1 Institute of Soil and Environmental Sciences, PMAS Arid Agricultural University, Rawalpindi 46000, Pakistan
2 Department of Horticulture, PMAS Arid Agriculture University, Rawalpindi 46000, Pakistan
3 Soil and Water Testing Laboratory, Data Ganj Baksh Road, Rawalpindi 46000, Pakistan
4 Department of Biosciences, COMSATS University, Islamabad 45550, Pakistan
* Correspondence: qaiser.hussain@uaar.edu.pk

† Presented at the 1st International Precision Agriculture Pakistan Conference 2022 (PAPC 2022)—Change the Culture of Agriculture, Rawalpindi, Pakistan, 22–24 September 2022.

Abstract: The diagnosis and recommendation integrated system (DRIS) approach assesses plants' nutrient status by comparing crop nutrient ratios with optimum values from a high-yielding group (DRIS norms). Furthermore, in this study, twenty-one grids were selected using GPS for soil sampling from a 1-hectare olive orchard field with a grid size of 20 × 20 m. Using standard laboratory procedures, soil samples were analyzed for macronutrients (N, P, and K ranging from 0.06–0.17%, 1.44–5.56%, and 31.94–120.32 mg kg^{-1}) and micronutrients (Fe, Cu, Mn, Zn and B ranging from 25.16–82.3, 1.09–1.8, 26.96–65.69, 0.01–4.5 and 0.48–1.06 mg kg^{-1}), respectively, and in plant samples for macronutrients (N, P and K ranging from 0.63–1.93, 0.01–0.16 and 0.75–1.37%) and micronutrients (Fe, Cu, Zn, Mn and B ranging from 152.5–621.5, 8.5–17, 18–34 and 43.5–113 mg kg^{-1}), respectively. The critical ranges were investigated in terms of their relationship to the population's yield level. DRIS norms derived from olives were able to detect nutrient deficiency and excess.

Keywords: nutrient indexing; DRIS; rainfed; macronutrients; yield

Citation: Khan, K.M.; Hussain, Q.; Akmal, M.; Khan, M.A.; Alvi, S.; Shakeel, T.; Manzoor, R.; Irfan, M. Nutrient Indexing of Different Olive Cultivars under Rainfed Conditions. *Environ. Sci. Proc.* **2022**, *23*, 11. https://doi.org/10.3390/environsciproc2022023011

Academic Editor: Saddam Hussain

Published: 19 December 2022

1. Introduction

The diagnosis and recommendation integrated system (DRIS) diagnoses nutrient deficiency by evaluating essential nutrient ratios in specific plant tissue rather than the absolute value of individual nutrient norms. Nutrient ratios in high-yield populations provide a reference value for comparing ratios found in low-yield populations of the same crop varieties. The DRIS method was developed to provide a valid diagnostic regardless of plant age, the cultivar being grown, prevailing conditions, changes in sampling method, or sampling time [1]. Hence, our study aimed to help understand the availability of soil nutrients and apply fertilizer based on recommendations to achieve the highest yields through the use of the nutrient index model (DRIS). Olive (*Olea europaea* L.) is considered the most important fruit tree due to its economic significance for oil production [2]. Olive is cultivated over an estimated area of 10.51 million hectares all around the globe, producing around 21.07 million tonnes of olive fruit [3].

2. Materials and Methods

The index soil samples collection was performed by selecting twenty-one grids of olive orchard fields by QGIS software and samples were collected at 0–30 cm depth from one hectare with a grid size of 20 × 20 m. Geospatial analysis was performed to quantify the degree of spatial dependence/variability by using ArcGIS software. Olive plant samples were also collected from different cultivars in each grid. The collected samples were oven dried at 65 °C for 24 h and after that grinding and storage were performed for further analysis. Phosphorus, K, Ca, Mg, B, Zn, Cu, Fe, and Mn in Olive plants were determined by the wet digestion method and the use of nitric acid and chloric acid at a 2:1 ratio [4]. The nutrients released were measured by ICP-OES [5]. The total nitrogen (N) in the plant samples was determined by the combustion method on an Elemetar Vario EL iii-combustion analyzer in CNS mode. Dry and ground plant samples (15 mg) were combusted with the tungsten oxide catalyst. The resulting gases passed through various heat and chemical traps [6].

3. Results and Discussion

3.1. GIS Mapping and Olive Orchard

Geospatial analysis was performed to quantify the degree of spatial dependence/variability by using ArcGIS software (Figure 1).

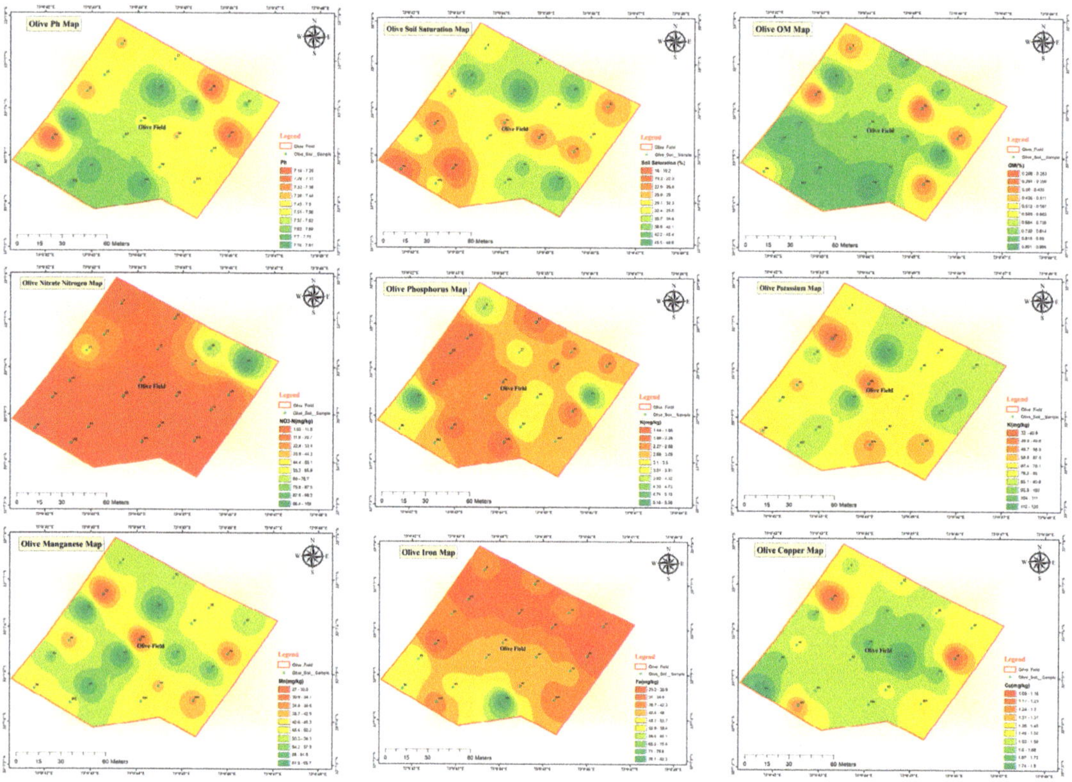

Figure 1. GIS maps showing status of different soil porperties and nutrients in the olive orchard.

3.2. DRIS Model

The diagnosis and recommendation integrated system (DRIS) is used for predicting whether a nutrient is present in sufficient quantity or not. Different DIRS norms developed in an olive orchard in Koont, Chakwal. Moreover, the nutrient status of olive trees is determined by critical level approach.

3.3. Nutrient Ratios for Low and High Yielders

Different nutrients for high- and low-yield ratios were varied, for example, the N/P ratio for high yielders was 12.91 while for low yielders its value was 13.33. Similarly, the N/K ratio values for high- and low-yielding populations were 1.13 and 1.12, respectively, and were not significantly different. N/Fe values for high and low yielders were 37.21 and 34.98, respectively, while for N/Cu these values were 925.99 and 1028.65, respectively. These were also not significantly different. Phosphorus to other nutrients ratios were as follows: P/N 0.10, P/K 38.9/48.4, P/Zn 3.35/3.32, P/Fe 81.6/95.8, P/Cu P/Mn 17.2/20.3 and P/B 26.3/28.8. For K to other nutrients were as follows: K/N 408/474, K/P 11.6/11.1, K/Zn 408/473, K/Fe 33.2/32.1, K/Cu 866/948, K/Mn 179/196 and K/B 290/177. For Fe, the ratios were as follows: Fe/N, 0.03, Fe/P, 0.37/0.40, Fe/K, 0.03/0.04, Fe/Zn 12.6/16.3, Fe/Cu 26.8/32.6, Fe/Mn 5.51/6.79 and Fe/B 9.18/6.28. Cu ratios with other nutrients were as follows: Cu/N 0.00/0.00, Cu/P 0.01/0.01, Cu/K 0.00/0.00, Cu/Zn 0.47/0.50, Cu/Fe 0.04/0.03, Cu/Mn 0.22/0.23 and Cu/B 0.34/0.20. Zn ratios with other nutrients were as follows: Zn/N 0.00/0.00, Zn/P 0.03/0.03, Zn/K 0.00/0.00, Zn/Fe 0.08/0.07, Zn/Cu 2.12/2.01, Zn/Mn 0.46/0.43 and Zn/B 0.72/0.41. For manganese, the ratios with other nutrients were as follows: Mn/N 0.01/0.01, Mn/P 0.06/0.06, Mn/K 0.01/0.01, Mn/Zn 2.30/2.44, Mn/Fe 0.19/0.16, Mn/Cu 4.87/4.92 and Mn/B 1.65/0.95. Lastly, for B with other nutrients, the ratios were as follows: B/N 0.00/0.01, B/P 0.04/0.07, B/K 0.00/0.01, B/Zn 1.47/3.20, B/Fe 0.13/0.22, B/Cu 3.08/6.28 and B/Mn 0.67/1.34.

DRIS norms, descriptive data and analysis of variance showed significant differences for boron and zinc contents, while the rest of the nutrient concentration as well as ratios remained statistically at par for high- and low-yielding populations (Table 1 and Figure 2). According to nutrient indexing of the DRIS model, boron and iron contents were sufficient in olive trees while nitrogen, phosphorus, potassium, zinc, copper, and manganese contents were moderate to highly deficient with zinc being the most deficient nutrient. Many studies have reported variations in the growth and yield of olive trees in response to their nutrient variations. These results are confirmed by [7] who reported that Arbequina showed higher growth and yield as compared to other olive cultivars as it was able to absorb more nutrients from the soil even under supersensitive conditions. In addition, [8] also reported results in favor of the present study and found that olive cultivars, especially Arbequina, were found to be more productive according to DRIS norms in long-term studies as compared to the studies that continued for shorter periods.

Table 1. DRIS nutrient index values according to Wadt methodology.

DRIS Nutrient Index	Index Value	DRIS Nutrient Index	Index Value
DRIS Index Nitrogen	−0.90647	DRIS Index Boron	25.11704
DRIS Index Phosphorus	−1.39084	DRIS Index Copper	−6.85626
DRIS Index Potassium	−1.18937	DRIS Index Iron	3.142205
DRIS Index Zinc	−10.8696	DRIS Index Manganese	−7.04673

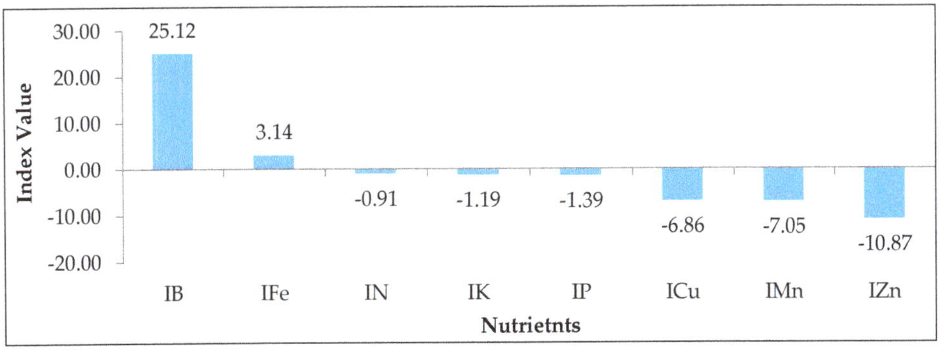

Figure 2. Nutrient index value for olive orchard.

4. Conclusions

Essential plant nutrients play an important role in increasing the production of olive fruit as well as the efficiency of olive oil extraction. These nutrients are taken up from the soil solution and then play their respective roles in plant metabolism after their assimilation into different plant parts as required. Olive plants lose these nutrients after the removal of fruits or by pruning. Management of these deficient nutrients via organic and inorganic fertilizers is essential for obtaining maximum fruit and oil production. DRIS model has provided a precise estimation of nutrient requirements and deficiency levels of different nutrients in an olive orchard that might prove to be very helpful in recommending the guidelines for fertilizer application and correcting (Figure 2).

Author Contributions: Conceptualization, K.M.K. and Q.H.; methodology, K.M.K. and Q.H.; formal analysis, K.M.K.; investigation, Q.H. and M.A.; resources, Q.H., T.S.; data curation, K.M.K., T.S.; writing—original draft preparation, K.M.K.; writing—review and editing, K.M.K., Q.H., M.A., T.S., M.A.K., M.I., S.A. and R.M.; visualization, Q.H.; supervision, Q.H. All authors have read and agreed to the published version of the manuscript.

Funding: This study was funded by Pilot Project for Data-Driven Smart Decision Platform (DDSDP) for Increased Agriculture Productivity.

Institutional Review Board Statement: Not Applicable.

Informed Consent Statement: Not Applicable.

Data Availability Statement: Not Applicable.

Acknowledgments: This study was funded by Pilot Project for Data-Driven Smart Decision Platform (DDSDP) for Increased Agriculture Productivity. Authors are also thankful to institute of Soil and Environmental Science for providing laboratory facilities during analysis.

Conflicts of Interest: The authors declare no conflict of interest.

References

1. Moreno, J.J.; Lucena, J.J.; Carpena, O. Effect of the iron supply on the nutrition of different citrus variety/rootstock combinations using DRIS. *J. Plant. Nutr.* **1996**, *19*, 689–704. [CrossRef]
2. Fernandez-Escobar, R.; Mundi-Prensa, J.D. *Olive Growing*, 1st ed.; Barranco, D., Fernández-Escobar, R., Rallo, L., Eds.; Australian Olive Association Ltd.: Pendle Hill, Australia, 2010; Volume 1, pp. 18–19.
3. Centeno, A.; García, J.M.; Gómez-del-Campo, M. Vegetative, productive and oil quality responses of 'Arbequina' and 'Picual' olive trees to foliar P and K application. *Grasas Y Aceites* **2020**, *71*, e356. [CrossRef]
4. Ryan, J.; Estefan, G.; Rashid, A. *Soil and Plant Analysis Laboratory Manual*, 2nd ed.; International Center for Agricultural Research in Dry Area ICARDA: Alepo, Syria, 2001; pp. 118–127.
5. Munter, R.C.; Grande, R.A. *Plant Analysis and Soil Extracts by ICP-Atomic Emission Spectrometry*, 1st ed.; Barnes, R.M., Ed.; Developments in Atomic Plasma Spectrochemical Analysis; Hcyden and Son, Ltd.: London, UK, 1981; pp. 653–672.
6. Kjeldahl, C. A new method for the determination of nitrogen in organic matter. *Anal Chem.* **1883**, *22*, 366. [CrossRef]
7. Gimenez, M.; Nieves, M.; Gimeno, H.; Tomé, J.M.; Nicolás, J.J.M. Nutritional diagnosis norms for three olive tree cultivars in superhigh-density orchards. *J. Agric. Nat. Resour.* **2021**, *48*, 34–44. [CrossRef]
8. Díez, C.M.; Moral, J.; Cabello, D.; Morello, P.; Rallo, L.; Barranco, D. Cultivar and tree density as key factors in the long-term performance of super high-density olive orchards. *Front. Plant Sci.* **2016**, *7*, 1226. [CrossRef] [PubMed]

environmental sciences proceedings

MDPI

Proceeding Paper

Precision Nitrogen Management for Cotton Using (GreenSeeker) Handheld Crop Sensors †

Hafiz Umar Farid [1,*], Zahid Mahmood Khan [1], Muhammad Naveed Anjum [2], Aamir Shakoor [1] and Huzaifa Shahzad Qureshi [1]

1 Department of Agricultural Engineering, Bahauddin Zakariya University, Multan 60800, Pakistan
2 Faculty of Agricultural Engineering & Technology, PMAS Arid Agriculture University, Rawalpindi 46000, Pakistan
* Correspondence: hufarid@bzu.edu.pk

† Presented at the 1st International Precision Agriculture Pakistan Conference 2022 (PAPC 2022)—Change the Culture of Agriculture, Rawalpindi, Pakistan, 22–24 September 2022.

Abstract: The precise monitoring of nitrogen (N) is an effective strategy for enhancing the crop yield per unit of land, but it involves field-level soil and crop data. The two years of experimental study were conducted during the cotton growing seasons of 2018 and 2019 at the Agriculture Research Farm of the Department of Agricultural Engineering, Bahauddin Zakariya University, Multan. The Nitrogen Fertilizer Optimization Algorithm (NFOA) was formulated based on the observed data for cotton lint yield (CLY) and GreenSeeker Normalized Difference Vegetation Index (GSNDVI) during the growing stages of cotton. The precision nitrogen application rate-based green seeker (PNAR) G.S for cotton was identified as 150-165 kg/ha. A linear relationship was observed between CLY (R2 = 0.80) for cotton with the GSNDVI. The average nitrogen requirement (N_{req}) using (PNAR) G.S was determined through the nitrogen fertilizer optimization algorithm (NFOA). The N_{req} was found to be 0.013 kg/kg for cotton. Precision N management originating from handheld crop sensors (GreenSeeker) may be helpful in decision-making for site-specific in-season N fertilizer management to enhance crop yield.

Keywords: Green-Seeker; cotton; NDVI; PNAR; NUE; NFOA

Citation: Farid, H.U.; Khan, Z.M.; Anjum, M.N.; Shakoor, A.; Qureshi, H.S. Precision Nitrogen Management for Cotton Using (GreenSeeker) Handheld Crop Sensors. *Environ. Sci. Proc.* **2022**, *23*, 12. https://doi.org/10.3390/environsciproc2022023012

Academic Editor: Tahir Iqbal

Published: 19 December 2022

1. Introduction

Precision N-management approaches are sustainable agricultural development practices. They have promoted overcoming the growing food demands and preventing declining natural food resources. Small landholders need to focus on these development approaches to maximize per unit land productivity. Hence, precision agriculture approaches are a fruitful method for upholding declining natural resources by applying inputs only at the right time when needed [1]. Additionally, Variable rate application (VRA) can calculate the N-fertilizer to resist the intra-field variabilities and to minimize the leaching of N-fertilizer because of over-application in the field. It has been reported that precision agriculture techniques, in combination with VRA, are the most effective way to make an N-management strategy based on spatial variability [2]. Moreover, N-management strategies can also be developed using a handheld sensor (Green-Seeker), providing real-time variable fertilizer rates over the field. These sensors are key tools for boosting crop yield per unit of land on less fertile soils. However, crop N sensors never provide the actual amount of required fertilizer; rather, the actual N status of plants is analyzed through vegetation indices through these crop N sensors. One of the most common and effective indices is the normalized difference vegetation indices (NDVI) that assist in generating in-season N requirement algorithms [3]. Depending upon the developed nitrogen fertilizer optimization algorithm (NFOA), plant N needs can be calculated easily to meet the requirement of the crop nutrient levels in the entire field [4]. Many researchers have reported the application of active crop sensors and handheld sensors (Green-Seeker) to determine the in-season nitrogen (N) application rates [5,6]. Therefore, the present research was conducted to develop the nitrogen fertilizer optimization algorithm (NFOA) and to adopt sensor-based algorithm approaches to formulate the precision input management strategy for cotton to achieve maximum productivity per unit of land.

2. Materials and Methods

The experimental field for the study was selected at the Agriculture Research Farm of the Department of Agricultural Engineering, Bahauddin Zakariya University, Multan (Figure 1). The research area was selected based on the easiness and timeliness of the soil collection, plant, and yield data, to ensure the proper analysis. The accessibility of all the field inputs (water, implements, and fertilizer) and laborers was ensured before site selection. The climatic conditions of Multan are arid, with sweltering summers and cold winters. Cotton seeds were sown using the "Chogga" method, which is adopted traditionally due to the unavailability of cotton planters. Cotton seeds sown at a rate of 2.02 kg/ha were applied to the field, and the variety of the seeds used in this experiment was "IUB-13". The nitrogen fertilizer optimization algorithm (NFOA) for the N-management strategy was formed with the help of the observed data of cotton crops during the 2018 and 2019 growing seasons. The Green-Seeker sensor was used to collect the NDVI data for wheat during its growing stages. The stages decided for the collection of the NDVI values were 3-feekes, 5-feekes, booting, and before the harvest of the crop. Moreover, the data for the observed cotton lint yield (CLY) were also used to determine their relationship with the GSNDVI. The relationship of CLY with GSNDVI for each of the growing stages of wheat was analyzed in MS Excel. The relationship that represents satisfactory performance was selected for the nitrogen fertilizer optimization algorithm (NFOA) to predict the in-season CLY. Moreover, these selected equations were used to determine the Final Precision Nitrogen Uptake (FPNU) and Early Season Precision Nitrogen Uptake (EPNU).

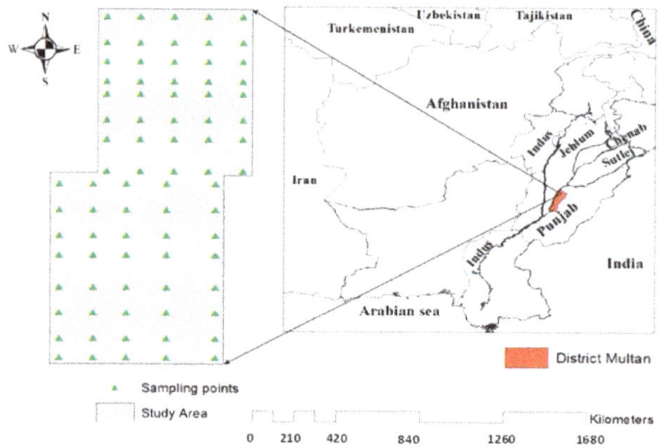

Figure 1. Location of the study area with a sampling position.

3. Results and Discussion

The data on the cotton lint yield (CLY) were plotted against the Green-Seaker-based in-season estimated yield (GSINSEY) (Figure 2a). This relationship provides a measure of the final plant N uptake (FPNU) by the crop for achieving a certain yield level and can prove to be useful in developing a nitrogen fertilizer optimization algorithm (NFOA) for precision N management. The relationship of INSEY vs. CLY performed better compared to other (days after planting) DAP days due to a higher coefficient of determination ($R^2 = 0.84$). Therefore, this relationship (CLY (kg/ha) = 804368 × GSINSEY − 979) was used to calculate the FPNU in NFOA development. It has been studied that final plant N uptake requires the accurate measurement of crop yield to establish a relationship of N requirement before generating a nitrogen fertilizer optimization algorithm (NFOA). The inaccurate estimation of yield may result in the over- or under-recommendation of N fertilizers [7]. The relationship between NDVI and in-season N requirement was developed for cotton crops and is shown in Figure 2. It analyzed that the coefficient of determination R^2 between the NDVI and N requirement was 0.82. The relationship between the NDVI and N requirement indicates a clear exponential regression to measure the in-season N application rate using Equation (1) as follows:

$$\mathbf{y = 141.09e^{-2.389x}} \quad (1)$$

where, y = total N (kg/ha) and x = in season NDVI value measured from sensor.

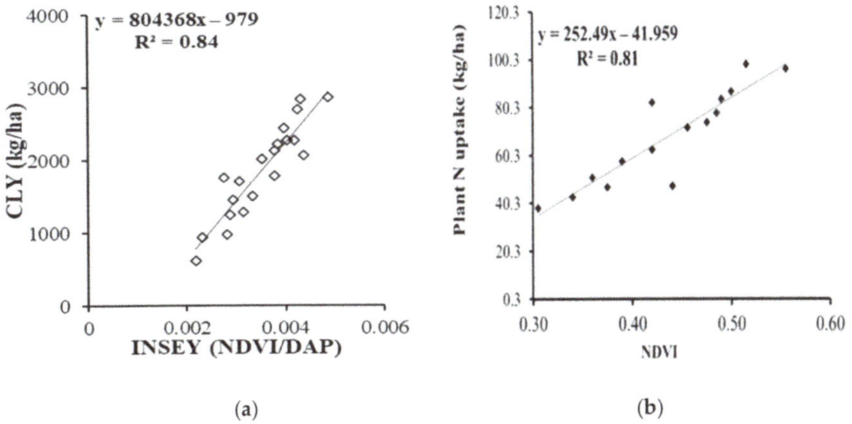

Figure 2. (**a**) Relationship of cotton lint yield (CLY) with INSEY (NDVI/DAP) data collected at different time intervals (**b**) plant N uptake and NDVI.

The relationship between sensor NDVI and cotton plant N uptake in the year 2018 is reported in Figure 2b. As crop plant N uptake is defined by the product of plant dry biomass and N content, it should be related to NDVI [5]. Therefore, the N application should be based on the relationship between sensor NDVI and N uptake by the plant. When the total N uptake by the cotton plant regressed against the NDVI readings, a strong linear model with a coefficient of determination (R^2 = 0.81) was observed. The EPA was determined using the following relationship of (EPNU (kg/ha) = 252.49 × GSNDVI − 41.95), which was analyzed between actual plant N uptake and the sensor NDVI values.

GreenSeeker-Based Topdressing N recommendation Algorithm for Cotton

Based on the above result determined in Figure 2a, the relationship of INSEY and CLY with higher (R^2) values was used to determine the final plant N uptake (FPNU). In the present study, FPNU was determined using the N requirement (N_{req}) and predicted CLY. N_{req} is defined as the amount of N needed to produce 1 kg of cotton. Therefore, the FPNU includes the N uptake by the cotton plant, and the N_{req} at different yield ranges was based on the results reported by [8]. According to the NFOA developed by [9], a precise in-season N application rate can be determined by taking the difference between the predicted EPNU and FPNU before topdressing divided by nitrogen use efficiency (NUE).

The following GreenSeeker (GS) nitrogen fertilizer optimization algorithm (NFOA) was proposed to determine the topdressing N fertilizer application rate for cotton crops:

- Predicting cotton yield using GS INSEY (NDVI-DAP) before topdressing N fertilizer application
    - Cotton lint yield CLY (kg/ha) = 804368 × (GSINSEY) − 979.

- The average N requirement (Nreq) of 0.013 kg/kg for cotton lint yield (CLY) of 2355 kg/ha was found based on the (GS) N application rate.
- Similarly, the average N requirement (Nreq) of 0.011 kg/kg for cotton lint yield (CLY) of 2335kg/ha was found based on the precision N application rate of soil analysis.

- Calculating final plant N uptake (FPNU) using predicted cotton yield and N_{req}, i.e.,
  - FPNU (kg N/ha) = CLY × N_{req}.
- Predicting early season plant N uptake (EPNU) using GSNDVI before topdressing
  - EPNU (kg N/ha) = 252.49 × GSNDVI − 41.95
- Determination of in-season topdressing N fertilizer requirement (NR)
  - NR (kg N/ha) = (FPNU − EPNU)/Nitrogen use efficiency (NUE)
  - NUE value was set to 40%.

4. Conclusions

The two years analysis of data for cotton indicated that a Green-Seeker handheld sensor could be used to determine the optimum N rate. The precision nitrogen management rate-based Green-Seeker $(PNAR)_{G.S}$ for cotton was identified as 150–165 kg/ha. A linear relation was also observed between CLY (R^2 = 0.80) for cotton with the GSNDVI. The average N requirement for cotton lint yield based on the nitrogen fertilizer optimization algorithm (NFOA) was 0.013 kg/kg, indicating the instant application of N fertilizer precisely by using handheld crop sensors. The performance evaluation of NFOA indicated satisfactory results based on their NSE, RMSE, NRMSE, and MAPE. Overall, the results showed that Green-Seeker-based N management strategies might increase crop yield.

Author Contributions: Conceptualization, H.U.F. and Z.M.K.; methodology, H.S.Q., H.U.F. and A.S.; investigation, H.U.F., M.N.A. and Z.M.K.; resources, H.U.F., M.N.A. and Z.M.K.; data curation, H.U.F., M.N.A., A.S., H.S.Q. and Z.M.K.; writing—original draft preparation, H.U.F., M.N.A., A.S., H.S.Q. and Z.M.K.; writing—review and editing, H.U.F., M.N.A., A.S., H.S.Q. and Z.M.K.; visualization, H.U.F., M.N.A., A.S., H.S.Q. and Z.M.K. All authors have read and agreed to the published version of the manuscript.

Funding: This research received no external funding.

Institutional Review Board Statement: Not applicable.

Informed Consent Statement: Not applicable.

Data Availability Statement: The data are available upon request from the corresponding author.

Acknowledgments: The authors would like to acknowledge the Higher Education Commission (HEC) for providing financial assistance under Project No. NRPU-8370 and Department of Agricultural Engineering, Bahauddin Zakariya University, Multan, for providing the facilities to conduct this research.

Conflicts of Interest: The authors declare no conflict of interest.

References

1. Kale, S.S.; Panzade, K.P.; Chavan, N.R. Modern Farming Methods: An Initiative towards Increasing the Food Modern Farming Methods: An Initiative towards Increasing the Food Productivity. *Food Sci. Rep.* **2020**, *1*, 34–36.
2. Farid, H.U.; Bakhsh, A.; Ahmad, N.; Ahmad, A.; Mahmood-Khan, Z. Delineating Site-specific management zones for precision agriculture. *J. Agric. Sci.* **2016**, *154*, 273–286. [CrossRef]
3. Tagarakis, A.C.; Ketterings, Q.M. Proximal sensor-based algorithm for variable rate nitrogen application in maize in northeast USA. *Comput. Electron. Agric.* **2018**, *145*, 373–378. [CrossRef]
4. Cao, Q.; Miao, Y.; Li, F.; Gao, X.; Liu, B.; Lu, D.; Chen, X. Developing a new Crop Circle active canopy sensor-based precision nitrogen management strategy for winter wheat in North China Plain. *Precis. Agric.* **2017**, *18*, 2–18. [CrossRef]
5. Ali, A.M.; Ibrahim, S.M.; Bijay-Singh. Wheat grain yield and nitrogen uptake prediction using atLeaf and GreenSeeker portable optical sensors at jointing growth stage. *Inform. Process. Agric.* **2020**, *7*, 375–383. [CrossRef]

6. Zhou, L.; Chen, G.; Miao, Y.; Zhang, H.; Chen, Z.; Xu, L.; Guo, L. Evaluating a Crop Circle active sensor-based in-season nitrogen management algorithm in different winter wheat cropping systems. *Adv. Anim. Biosci.* **2017**, *8*, 364–367. [CrossRef]
7. Yao, Y.; Miao, Y.; Huang, S.; Gao, L.; Ma, X.; Zhao, G.; Jiang, R.; Chen, X.; Zhang, F.; Yu, K.; et al. Active canopy sensor-based precision N management strategy for rice. *Agron. Sustain. Dev.* **2012**, *32*, 925–933. [CrossRef]
8. Yue, S.; Meng, Q.; Zhao, R.; Ye, Y.; Zhang, F.; Cui, Z.; Chen, X. Change in nitrogen requirement with increasing grain yield for winter wheat. *Agron. J.* **2012**, *104*, 1687–1693. [CrossRef]
9. Lukina, E.V.; Freeman, K.W.; Wynn, K.J.; Thomason, W.E.; Mullen, R.W.; Stone, M.L.; Solie, J.B.; Klatt, A.R.; Johnson, G.V.; Elliott, R.L.; et al. Nitrogen fertilization optimization algorithm based on in-season estimates of yield and plant nitrogen uptake. *J. Plant Nutr.* **2001**, *24*, 885–898. [CrossRef]

environmental sciences proceedings

MDPI

Proceeding Paper

Optimization of Intelligent Irrigation Systems for Smart Farming Using Multi-Spectral Unmanned Aerial Vehicle and Digital Twins Modeling †

Muhammad Awais [1], Wei Li [1,*], Haoming Li [1], Muhammad Jehanzeb Masud Cheema [2,3], Saddam Hussain [3,4,5] and Chenchen Liu [1]

1 Research Center of Fluid Machinery Engineering & Technology, Jiangsu University, Zhenjiang 212013, China
2 Department of Land and Water Conservation Engineering, PMAS Arid Agricultural University, Rawalpindi 46000, Pakistan
3 National Center of Industrial Biotechnology (NCIB), PMAS Arid Agricultural University, Rawalpindi 46000, Pakistan
4 Department of Irrigation and Drainage, University of Agriculture Faisalabad, Faisalabad 38000, Pakistan
5 Department of Agricultural and Biological Engineering, University of California (UC Davis), Davis, CA 95616, USA
* Correspondence: lwjiangda@ujs.edu.cn; Tel.: +86-186-5285-0503

† Presented at the 1st International Precision Agriculture Pakistan Conference 2022 (PAPC 2022)—Change the Culture of Agriculture, Rawalpindi, Pakistan, 22–24 September 2022.

Abstract: This research presents the new techniques and practical experiences of using unmanned aerial vehicles (UAVs) precision agriculture mapping. UAV-based remote sensing systems should be cost-effective, fast-producing, have high geometric accuracy, and be simple to operate by local staff. This work aims to: (1) precisely use high-resolution UAV thermal multi-spectral sensors and machine learning approaches to reliably assess crop water status on a field scale; (2) capture on-field images for quantitative study from the multi-spectral sensors; (3) establish workflows for digital agriculture applications; (4) interpret the intelligent irrigation decision model using UAV indices, maps, and multi-source heterogeneous data integration. This research gives us new methods to set an intelligent method for precision agriculture, which greatly improves the level of agricultural intelligence.

Keywords: UAV application; soil monitoring; infrared image processing; principal component analysis; digital twin

Citation: Awais, M.; Li, W.; Li, H.; Cheema, M.J.M.; Hussain, S.; Liu, C. Optimization of Intelligent Irrigation Systems for Smart Farming Using Multi-Spectral Unmanned Aerial Vehicle and Digital Twins Modeling. *Environ. Sci. Proc.* **2022**, *23*, 13. https://doi.org/10.3390/environsciproc2022023013

Academic Editor: Muhammad Naveed Tahir

Published: 19 December 2022

1. Introduction

The growing global population necessitates an increase in food production, which consumes around 85% of the freshwater resources available [1]. The biggest hurdle preventing China and other emerging countries from achieving long-term sustainable development is a lack of water, and water crises will become the biggest concern for the next 10 years [2]. At present, digital twin technology, one of the top ten key technologies for the future, has been applied to the field of smart agricultural irrigation [3,4]. Soil with inadequate drainage capacity and a hard layer is not suitable for rice–wheat production [5]. The overarching aim of this study is to: (1) precisely use high-resolution UAV thermal multi-spectral sensors and machine learning approaches to reliably assess crop water status on a field scale; (2) capture on-field images for quantitative study from the multi-spectral sensors; (3) establish workflows for digital agriculture applications; (4) interpret the intelligent irrigation decision model using UAV indices, maps, and multi-source heterogeneous data integration. This experiment was performed in a tea field located in Jurong, China. The outcomes of this research will have a great benefit for both farmers and the industry.

2. Materials and Methods

2.1. Description of the Test Area

This experiment was performed on cultivated land located in the Maoshan Tea Garden experimental zone in Jurong City (32°1′00″ N, 119°4′00″ E), Jiangsu Province, China (Figure 1). The texture of the field soil is silty loam. Most of the instruments were deployed in the center of the study field on a flux tower to ensure that the prevailing wind direction had the most significant footprint. The tea plants (Camellia Sinensis) were six years old, with row and plant spacing of 1.5 and m, respectively.

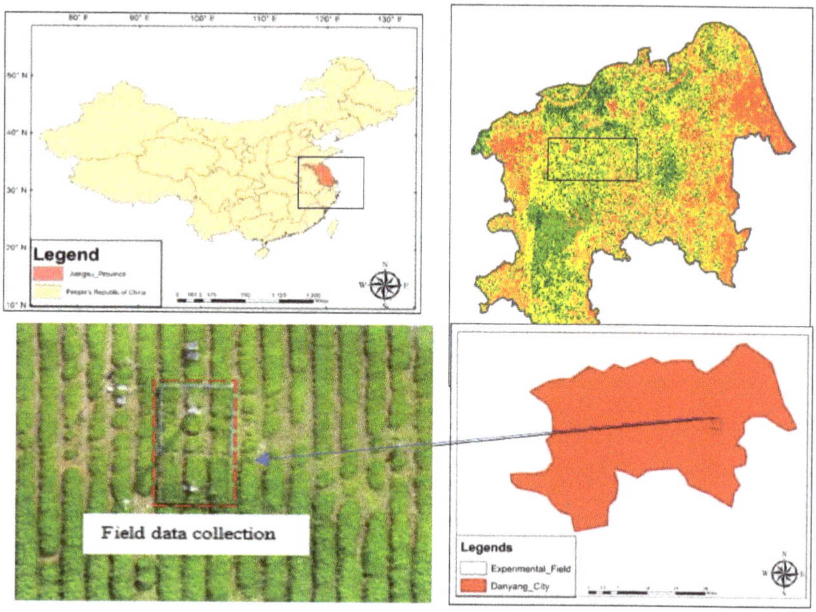

Figure 1. Study map of the experimental site.

2.2. Airborne Image Acquisition

The digital and thermal photos were collected using a quad-rotor UAV equipped with a multi-spectral sensor to capture spectral photos (Figure 2). During crop growth, we performed an airborne campaign to gather photos at various times of the day (9.00, 11.00, and 14.00 h.). Throughout the experimental field, several arbitrary GCPs (ground control points) were measured, and coordinates were calculated with a total precision of 0.1 m. For the orthomosaic map and picture pre-processing, a Pix-4D mapper and DJI Terra were applied. This program was created primarily for photogrammetry and computer visualization techniques to handle UAV images.

(**a**) (**b**)

Figure 2. Types of UAVs and sensors used in this study: (**a**) quad-rotor UAV with RGB sensor, DJI Phantom 4 RTK, (**b**) flying operations.

2.3. Intelligent Decision-Making Irrigation Systems

Figure 3 shows the design framework of the system, which deeply integrates the digital twin, the internet of things, big data, wireless transmission technology, cloud computing, and automatic control technology to build a physical layer, a data acquisition layer, a twin model layer, a functional layer, and an application layer. In addition, it is necessary to build the hardware perception and control system of the digital twin irrigation system from the perspective of the system level. With the help of various types of sensors and electrical control methods, the interconnection and intercommunication of various types of irrigation equipment in farmland can be realized, so as to carry out unified information operation, maintenance, and control.

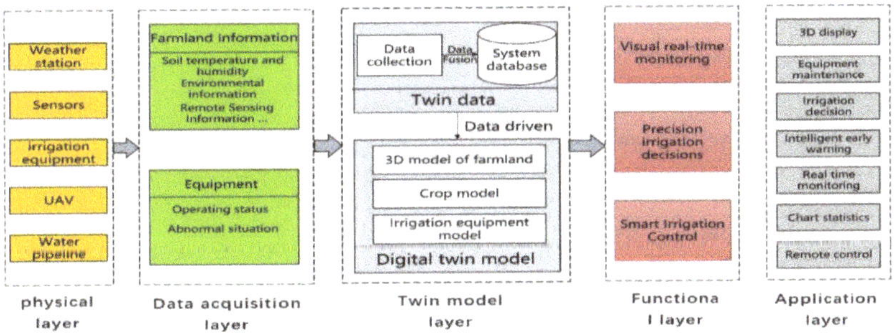

Figure 3. Block diagram for an intelligent irrigation system using a digital twin.

3. Results and Discussion

In this paper, a complete field automatic irrigation control system was built through the whole system design and the selection of the system hardware. The irrigation supervisory control system was tested (Figure 4). The pipeline used in this test was the PVC pipeline. The diameter was 20 mm, and the distance between the upstream and downstream probes was 4.05 mm, according to the calculation. The main tests were: the reliability test of the circuit hardware, the stability test of the wireless network communication, the security test of the power supply system, and the overall operation test. The wireless network communication status test is mainly about the communication distance and the networking stability of the communication module. The power supply system test is mainly about the safety and stability of the battery power supply, as shown in Figure 4 below. After the test is completed in the laboratory, the equipment is installed in the tea garden irrigation system for field application.

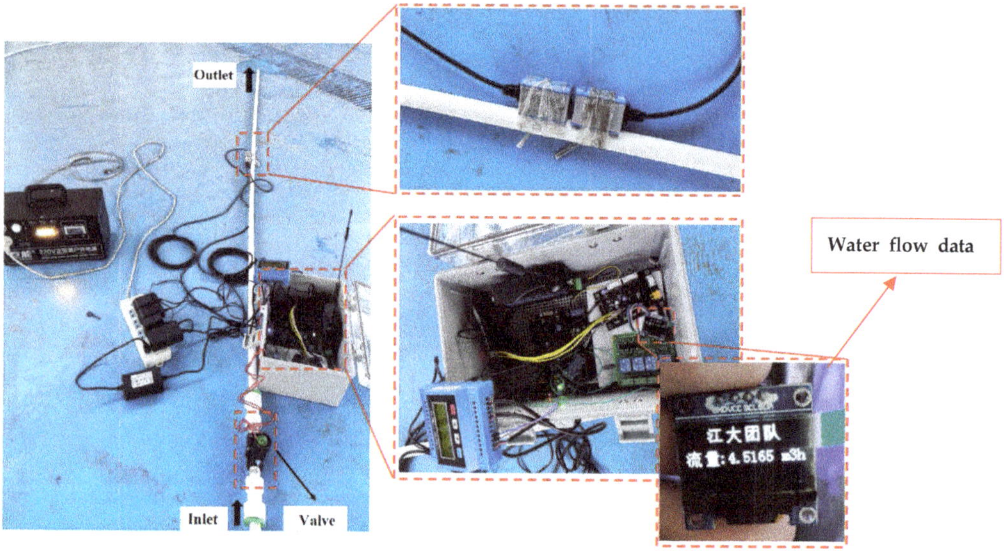

Figure 4. Spot application test.

Author Contributions: Methodology, M.A.; Investigation, W.L.; Conceptualization, H.L.; Writing—review & editing, M.J.M.C.; Data curation, S.H.; Writing original draft, C.L. All authors have read and agreed to the published version of the manuscript.

Funding: We acknowledge support from the Jiangsu Funding Program for Excellent Postdoctoral Talent (2022ZB667), "Belt and Road" Innovation Cooperation Project of Jiangsu Province (No.BZ2020068), Independent Innovation Fund Project of Agricultural Science and Technology in Jiangsu Province (No.CX (20)2037), and the Synergistic Innovation Center of Jiangsu Modern Agricultural Equipment and Technology (No.4091600014).

Institutional Review Board Statement: Not applicable.

Informed Consent Statement: Not applicable.

Data Availability Statement: The data presented in this study are available on request from the corresponding author. The data are not publicly available due to privacy restrictions.

Conflicts of Interest: The authors declare no conflict of interest.

References

1. Talaviya, T.; Shah, D.; Patel, N.; Yagnik, H.; Shah, M. Implementation of artificial intelligence in agriculture for optimisation of irrigation and application of pesticides and herbicides. *Artif. Intell. Agric.* **2020**, *4*, 58–73. [CrossRef]
2. Yang, Y. Dynamic Relationship of Urban and Rural Water Shortage Risks Based on the Economy–Society–Environment Perspective. *Agriculture* **2022**, *12*, 148. [CrossRef]
3. Karakoçak, B.B.; Yenigun, O.; Toraman, R.T. An integrated approach to water management in Kayseri: Rainwater collection and use in an amusement park. *Water Sci. Technol.* **2013**, *67*, 1137–1143. [CrossRef] [PubMed]
4. Wang, J.; Li, Y.; Huang, J.; Yan, T.; Sun, T. Growing water scarcity, food security and government responses in China. *Glob. Food Secur.* **2017**, *14*, 9–17. [CrossRef]
5. Song, X.; Wu, F.; Lu, X.; Yang, T.; Ju, C.; Sun, C.; Liu, T. The Classification of Farming Progress in Rice–Wheat Rotation Fields Based on UAV RGB Images and the Regional Mean Model. *Agriculture* **2022**, *12*, 124. [CrossRef]

environmental sciences proceedings

MDPI

Proceeding Paper

Impact of Integrated Nutrient Management on Yield of Different Varieties of Oat †

Lubna Anjum [1,*], Abdul Rehman [1], Muhammad Rizwan [2], Saddam Hussain [1,3,4] and Muhammad Sohail Waqas [5]

1 Department of Irrigation and Drainage, University of Agriculture Faisalabad, Faisalabad 38000, Pakistan
2 Department of Civil Engineering, Swedish College of Engineering and Technology, Rahim Yar Khan 64200, Pakistan
3 National Center for Industrial Biotechnology (NCIB), PMAS-Arid Agriculture University, Rawalpindi 46000, Pakistan
4 Department of Agricultural and Biological Engineering, University of California (UC Davis), Davis, CA 95616, USA
5 Soil Conservation Group, Agriculture Department (Field Wing), Government of the Punjab, Rawalpindi 46000, Pakistan
* Correspondence: lubnaanjum01@gmail.com
† Presented at the 1st International Precision Agriculture Pakistan Conference 2022 (PAPC 2022)—Change the Culture of Agriculture, Rawalpindi, Pakistan, 22–24 September 2022.

Abstract: Oat is an essential Rabi crop commonly used as green fodder. The nutritional requirements of oats are higher than those of other Rabi fodder crops. Higher doses of inorganic fertilizers are required to meet this demand, which is not economical for fodder production. This study evaluates the best economical dose of integrated nutrient management for attaining higher yield and better nutritional quality of oat. The experiment was conducted on the Agronomy Research Farm, University of Agriculture, Faisalabad, Punjab, Pakistan. The seed varieties used in this experiment were Oat-2011 Sargodha and Oat-F-411 (2021) Faisalabad. The experimental treatments were organized in factorial Randomized Complete Block Design (RCBD) arrangements, with each treatment replicated three times. The results revealed that treatment (T4) 50% RDF + 5 t/ha. Farmyard Manure + seed inoculation azotobacter was found to be the most appropriate than all other treatments.

Keywords: fodder; nutritional quality; Oat

Citation: Anjum, L.; Rehman, A.; Rizwan, M.; Hussain, S.; Waqas, M.S. Impact of Integrated Nutrient Management on Yield of Different Varieties of Oat. *Environ. Sci. Proc.* **2022**, 23, 14. https://doi.org/10.3390/environsciproc2022023014

Academic Editor: Tahir Iqbal

Published: 19 December 2022

1. Introduction

Agriculture is a vital component of Pakistan's economy. Pakistan's agricultural contribution to the gross domestic product is 21%, with a 2.7% annual growth rate [1]. Under irrigated conditions, it is an important winter fodder crop, either as a sole crop or in combination with barseem. Oat is a single-cut crop that provides fodder for a shorter period of time than barseem. In Pakistan, oat is grown under both rainfed and irrigated conditions. Oat is one of the most important cereal fodder crops grown in Pakistan during the winter [2]. The majority of oat fodder is fed in its green state, but any excess is dried or fermented into hay or silage to store for use during times of fodder shortages. It is a favorite food source for all types of animals, and its straw is superior to that of wheat and barley in both quality and pliability. Oat grain is also a good source of food for dairy cows, pigs, chickens, and other young animals that are used for breeding [3].

Traditional fodder crops in the area can feed half as many animals per unit of land, while improved oat types have the ability to produce three times as much green fodder as traditional fodder crops, which would be between 60 and 80 tonnes per acre [4]. Crop development and normal growth are primarily determined by the availability of irrigation water. Crop water requirements are related to the moisture-sensitive period. The authors of [5] demonstrated that during particular phases of plant development, the plant may be or appear to be more sensitive to changes in the amount of moisture present in the soil than it is during other stages of plant development. It can grow in a variety of soil types, altitudes, and rainfall conditions. It can withstand wet conditions better than most other cereals. It grows best in temperate and cool sub-tropical climates. During the four-month period, a well-distributed rainfall of 500 to 600 mm and an optimum temperature range of (18–35) °C are sufficient to meet its necessities as a fodder crop [6]. As a result, scientific society was driven to investigate the concept of integrated nutrient usage. Integrated nutrient management has good prospects, not only for ensuring high productivity but also for preventing soil degradation. Continuous use of large amounts of chemical fertilizers has had a negative effect, resulting in a decline in productivity due to a lack of one or more micronutrients. It also had a negative impact on soil health, resulting in several other issues.

Furthermore, to reduce the environmental impacts and increases in prices connected with chemical fertilizers, organic sources of nutrients are now emerging as attractive choices that can be used in conjunction with inorganic fertilizers because they are naturally balanced [7]. Integrated nutrient management (INM) using a combination of organic fertilizers and inorganic fertilizers may benefit soil characteristics and crop output. This might be performed in a sustainable way that does not compromise soil health, environmental safety, or other natural resources. Aside from this, INM procedures assist in lowering production costs and increasing farmer returns to evaluate the best economical dose of integrated nutrient management for attaining higher yield and better nutritional quality of oat. Keeping the above situation in view, the present study plans to investigate the impact of integrated nutrient management on the yield of different varieties of oat.

2. Methodology

The experimental site with geographical coordinates of 31°43′ N and 73°07′ E was conducted on oat crop during the year 2021–2022 at the agronomy-farm university of agriculture Faisalabad, Punjab, Pakistan. The experimental site is in Rachna Doab [8]. Faisalabad is located in the semi-arid region of Punjab, with humid summers and dry winters. Faisalabad's climate is classified as a hot desert. Summer temperatures can reach 50 °C as shown in Figure 1.

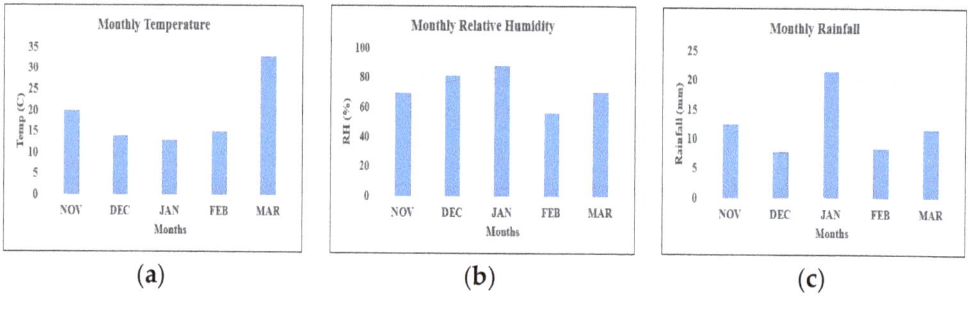

Figure 1. (**a**) Average monthly temperature; (**b**) monthly relative humidity; (**c**) average monthly rainfall.

2.1. Soil Sample Analysis

Soil samples were collected from a crop field with flood irrigation fields. Soil samples were collected from various points throughout the field to create a composite sample before sowing and after harvesting to get the status of soil (Figure 2).

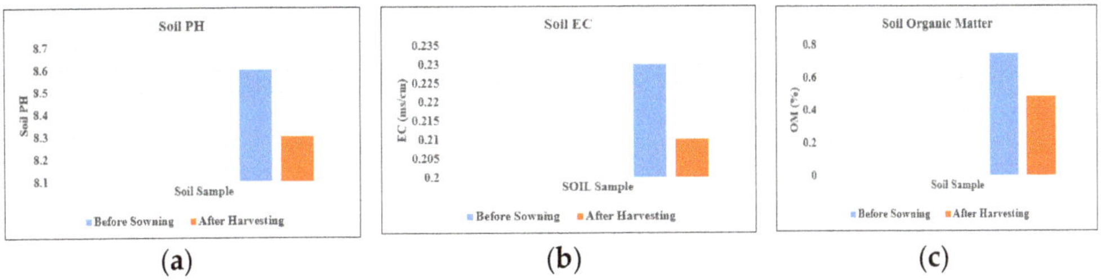

Figure 2. (**a**) Soil PH; (**b**) soil EC; (**c**) soil organic matter.

2.2. Treatments: Factor A: Varieties

V1: Oat-2011 (2011) SGD; V2: Oat-F-411 (2021) FSD
Factor B: Integrated Nutrient Management (INM)
T1: Recommended dose of fertilizer (RDF) N:P:K = 150:85:62 kg ha^{-1} (control)
T2: RDF + 5 t ha^{-1} Farmyard Manure
T3: RDF + 5 t ha^{-1} Poultry Manure
T4: 50% RDF + 5 ton ha^{-1} Farmyard Manure + (seed inoculation with azotobacter)
T5: 50% RDF + 5 ton ha^{-1} Poultry Manure + (seed inoculation with azotobacter)
T6: 25% RDF + 5 ton ha^{-1} Farmyard Manure + (seed inoculation with azotobacter)
T7: 25% RDF + 5 ton ha^{-1} Poultry Manure + (seed inoculation with azotobacter)

3. Results

Effect of (INM) on Oat Varieties with Different Treatments

The cultivars were quite different from one another in many respects. The cultivar Faisalabad F-411 had a significant effect on germination count (m^2), plant height (cm), number of tillers (m^2), leaf area per tiller (cm^2), number of leaves per tiller, green forage yield (t ha^{-1}), dry matter yield (t ha^{-1}), dry matter, crude protein, crude fiber, total ash and ether-extractable fat (%) under INM in all treatments of F 411 (2021) as compared to Oat 2011 SGD (Figures 3 and 4). The maximum germination count (208.4 m^2), plant height (144.7 cm), number of tillers (780.35 m^2), no. of leaf per tiller (6.49), leaf area per tiller (202.43 cm^2), leaf-to-stem ratio (0.414), green forage yield (72.1 t ha), dry matter yield (11.8 t ha), dry matter (16.31%) crude protein (10.22%), total ash (34.6%) and ether-extractable fat (4.5%) were recorded from the cultivar Faisalabad F-411.

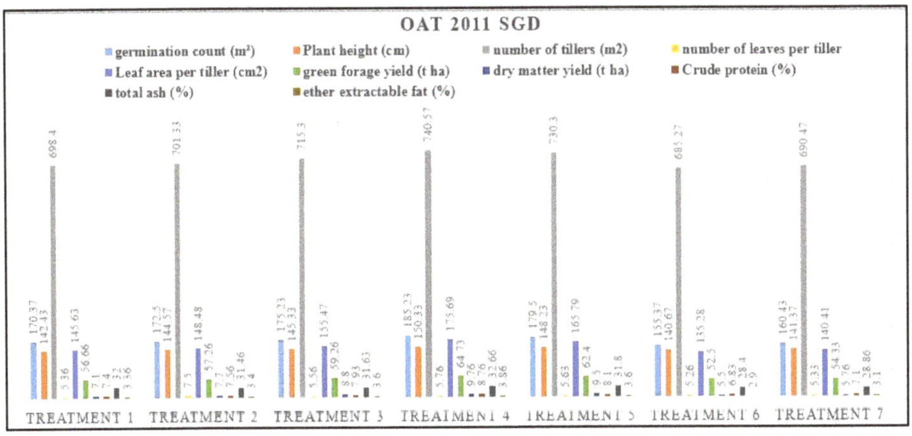

Figure 3. Effect of (INM) on oat cultivar Sargodha 2011.

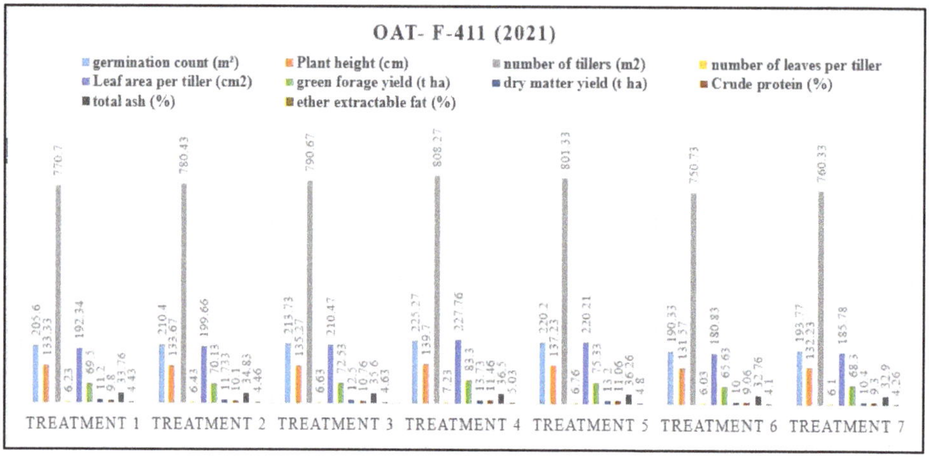

Figure 4. Effect of (INM) on oat cultivar F-411 Faisalabad.

4. Conclusions

From the results of the experiment and keeping overall performance in view, it can be concluded that treatment T_4 50% RDF + 5 t ha^{-1} Farmyard Manure + seed inoculation azotobacter was found to be the most appropriate of all the treatments studied in the experiment for exploiting the yield potential of oat (*Avena sativa* L.), and cultivar Faisalabad F-411 gave more forage yield of good quality under Faisalabad conditions.

Author Contributions: L.A. and A.R. gathered and processed the data set and performed the experiments. L.A. and M.R. supervised the experiments. L.A., A.R., M.R., S.H. and M.S.W. contributed to analyzing the results and writing the manuscript. All authors have read and agreed to the published version of the manuscript.

Funding: This research received no external funding.

Institutional Review Board Statement: Not applicable.

Informed Consent Statement: Not applicable.

Data Availability Statement: The data is available upon request from the corresponding author.

Conflicts of Interest: The authors declare no conflict of interest.

References

1. Azam, A.; Shafique, M. Agriculture in Pakistan and its Impact on Economy—A Review. *Int. J. Adv. Sci. Technol.* **2017**, *103*, 47–60. [CrossRef]
2. Muhammed, D.; Serkan, A. In Intercropping of legumes with Cereal crops in particular with the Perennials, Perennial Crops for Food Security. In Proceedings of the FAO Expert Workshop, Rome, Italy, 28–30 August 2018; p. 221.
3. uz Zaman, G.; Hussain, M.N.; Aziz, A. Performance of high yielding oat varieties under agro)-e cological conditions. *J. Agric. Res.* **2006**, *44*.
4. Haqqani, A.; Ali, Z.; Shafique, S.; Zahid, S.; Bakhsh, A. Oats: A fodder of winter lean period. *Agridigest* **2003**, *23*, 15–23.
5. Ghafoor, R.; Akram, N.A.; Rashid, M.; Ashraf, M.; Iqbal, M.; Lixin, Z. Exogenously applied proline induced changes in key anatomical features and physio-biochemical attributes in water stressed oat (*Avena sativa* L.) plants. *Physiol. Mol. Biol. Plants* **2019**, *25*, 1121–1135. [CrossRef] [PubMed]
6. Dost, M. *Fodder Success Story: Improved Fodder Crop Production in the Northern Areas of Pakistan*; Food and Agriculture Organization: Rome, Italy, 2001.
7. Mahato, M.; Biswas, S.; Dutta, D. Effect of Integrated Nutrient Management on Growth, Yield and Economics of Hybrid Maize (*Zea mays* L.). *Curr. J. Appl. Sci. Technol.* **2020**, *39*, 78–86. [CrossRef]
8. Jehangir, W.A.; Qureshi, A.S.; Ali, N. *Conjunctive Water Management in the Rechna Doab: An Overview of Resources and Issues*; International Water Management Institute: Colombo, Sri Lanka, 2002.

environmental sciences proceedings

MDPI

Proceeding Paper

Hybrid Renewable Energy Sources (Solar and Wind) Potential and Its Application for Sustainable Agriculture in Pakistan: A Case Study of Potohar Plateau †

Muhammad Saadullah Khan [1,*], Saddam Hussain [1,2,3], Muhammad Jehanzeb Masud Cheema [1,4], Tahir Iqbal [4], Shoaib Rashid Saleem [4] and Muhammad Sohail Waqas [5]

1 National Center for Industrial Biotechnology, PMAS-Arid Agriculture University, Rawalpindi 46000, Pakistan
2 Department of Irrigation and Drainage, University of Agriculture Faisalabad, Faisalabad 38000, Pakistan
3 Department of Agricultural and Biological Engineering, University of California (UC Davis), Davis, CA 95616, USA
4 Faculty of Agricultural Engineering and Technology, PMAS-Arid Agriculture University, Rawalpindi 46000, Pakistan
5 Soil Conservation Group, Agriculture Department (Field Wing), Government of the Punjab, Rawalpindi 46000, Pakistan
* Correspondence: saadkhan32@uaar.edu.pk
† Presented at the 1st International Precision Agriculture Pakistan Conference 2022 (PAPC 2022)—Change the Culture of Agriculture, Rawalpindi, Pakistan, 22–24 September 2022.

Abstract: The agriculture sector in Pakistan has a significant amount of potential in terms of solar, wind, geothermal, and biomass energy production and consumption. Most of the farm machinery on agriculture farms runs on high-cost fossil fuels and is also our source of greenhouse gas emissions (GHG). Utilizing renewable energy (RE) technology in the agriculture sector will reduce the cost of agricultural farming and will lower GHG emissions. The objective of this study is to determine and compare the solar and wind energy potential for the different districts in the Potohar region of Pakistan. The solar and wind energy data are obtained from HOMER Pro, RETScreen, and the weather station of the National Center of Industrial Biotechnology (NCIB), PMAS-Arid Agriculture University Rawalpindi. The results obtained are analyzed and compared which depicts that the Potohar region of Punjab, Pakistan has a significant amount of solar potential for its application in agriculture. This article gives an overview of renewable solar and wind energy potential and its applications for farmers and ranchers to make RE a rising source of energy and rural income in Pakistan.

Keywords: renewable energy; sustainable agriculture; climate change; agriculture farm; solar energy; wind energy; precision agriculture; sustainable production

Citation: Khan, M.S.; Hussain, S.; Cheema, M.J.M.; Iqbal, T.; Saleem, S.R.; Waqas, M.S. Hybrid Renewable Energy Sources (Solar and Wind) Potential and Its Application for Sustainable Agriculture in Pakistan: A Case Study of Potohar Plateau. *Environ. Sci. Proc.* **2022**, *23*, 15. https://doi.org/10.3390/environsciproc2022023015

Academic Editor: Qamar Zaman

Published: 20 December 2022

1. Introduction

According to the economic survey of Pakistan 2022 [1] report, Pakistan's overall energy generation capacity climbed 11.5% between April and July 2022, rising to 41,557 Megawatt (MW) from 37,261 MW during the same period in the previous fiscal year. Pakistan is reportedly experiencing an above 5000 MW electricity shortfall and is having difficulty harnessing its energy resources. Electricity consumption in the Pakistan agricultural sector has marginally grown from 8.9% to 9% in Fiscal Year (FY) 2022.

Nowadays, the agriculture sector's needs are geared toward modernization and efficiency to compete in a globalized market, and one of the challenges that must be addressed is escalating energy prices. According to the Food and Agriculture Organization (FAO), agri-food chain systems now utilize 30% of the world's energy production, with transportation, processing, packing, shipping, storage, and marketing consuming around 70% of the energy [2]. By 2050, renewable energy would account for 63% of all primary energy supply, up from 14% in 2015. In comparison to former years, this translates to an average yearly annual growth of 1.4%, which is a six-fold increase. The share of fossil fuels would also decrease, from 86% to 37% [3]. Pakistan's maximum electricity is generated from thermal energy comprising 61% of the generation capacity while the remaining energy is generated through hydel 24%, nuclear 12%, and a very small amount by RE sources 3% [1].

Pakistan's present energy environment has failed to meet the energy demands of domestic, industrial, agricultural, and different sectors. In these circumstances, it is necessary to rely increasingly on renewable energy supplies, particularly in remote places without access to the national grid. The landowners and farmers should be encouraged to adopt RE technologies through incentives and subsidies. Utilizing renewable energy sources for agricultural purposes could provide farmers with a steady source of income. The schematic diagram of hybrid RE systems that could be used in agricultural farms has been illustrated in Figure 1.

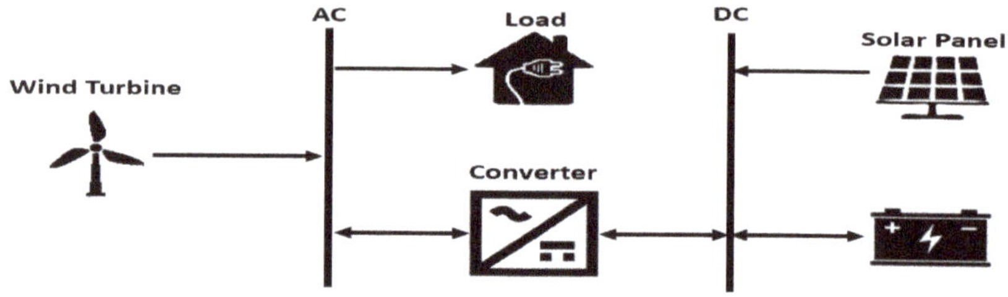

Figure 1. Schematic diagram of the off-grid hybrid renewable energy system.

2. Materials and Methods

2.1. Study Area

The Potohar Plateau comprises the districts Jhelum, Attock, Chakwal, and Rawalpindi. There are 2.2 mha of land on the Potohar Plateau [4]. The location map of the study area is shown in Figure 2 and the geographical coordinates of the Potohar Plateau districts.

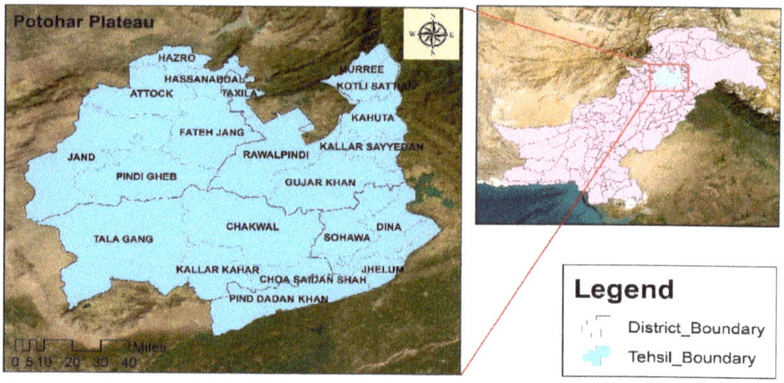

Figure 2. Location map of the study area (Potohar Plateau).

2.2. Methods

Hybrid energy modeling tools such as HOMER Pro [5], and RETScreen [6] have been used to obtain the solar and wind energy data for the study areas. HOMER Pro fetches the data for solar irradiance over 22 years from July 1983 to June 2005 from NASA's prediction of worldwide energy resources [7]. Solar irradiance data has also been obtained from the National Renewable Energy Laboratory (NREL) database [8]. The wind energy potential and temperature data are obtained from NASA for 30 years from January 1984 to December 2013. The real-time measured data of wind speed and temperature from March 2022 to October 2022 have been obtained from the weather station of the National Center of Industrial Biotechnology (NCIB), PMAS-Arid Agriculture University Rawalpindi, Pakistan. The solar and wind energy potential of the study areas were compared and analyzed for installation of hybrid RE systems for its application in precision agriculture.

3. Results and Discussion

As represented in Figure 3a, the average annual solar irradiance data obtained from HOMER Pro and NREL were compared. The comparison shows that all the districts of the Potohar plateau have sufficient solar energy potential for its application in smart agriculture farms. Figure 3b shows the comparison of average annual wind speed data obtained from HOMER Pro and RETScreen software. It is quite clear that study areas have comparatively low wind speed potential for the installation of high speed, and high-capacity wind turbines. However, low wind speed turbines such as EOCYCLE E010, having cut-in wind speed of 2.75 m/s and nominal capacity of 10 kW, could be installed in such sites. The nominal generation capacity of these wind turbines is sufficient for small farmers in rural farm locations in remote areas. Figure 4 represents a comparative analysis of wind speed potential data for the Rawalpindi district obtained from HOMER Pro, RETScreen, and real-time data obtained from the weather station of NCIB.

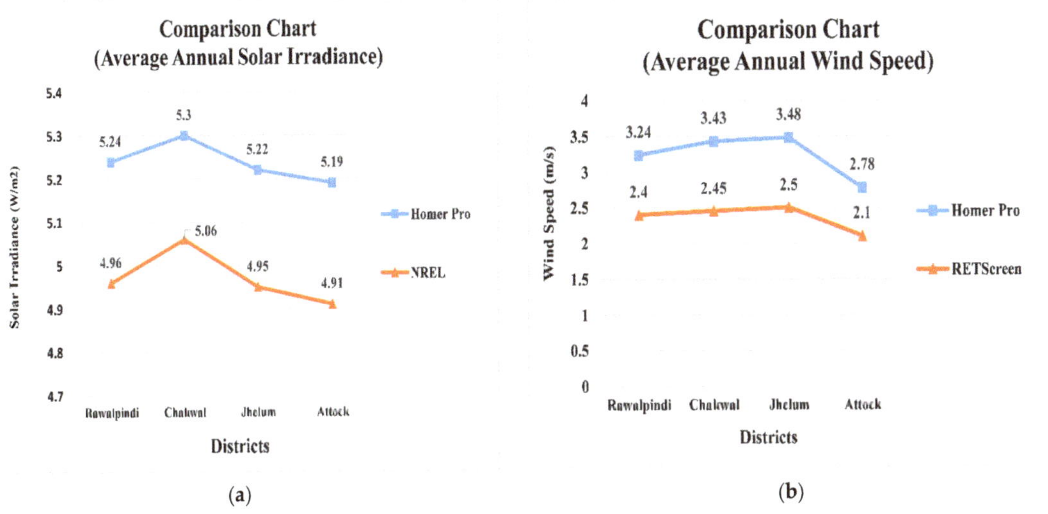

Figure 3. Comparison chart of average annual solar irradiance and wind speed of study area (**a**) average annual solar irradiance (**b**) average annual wind speed.

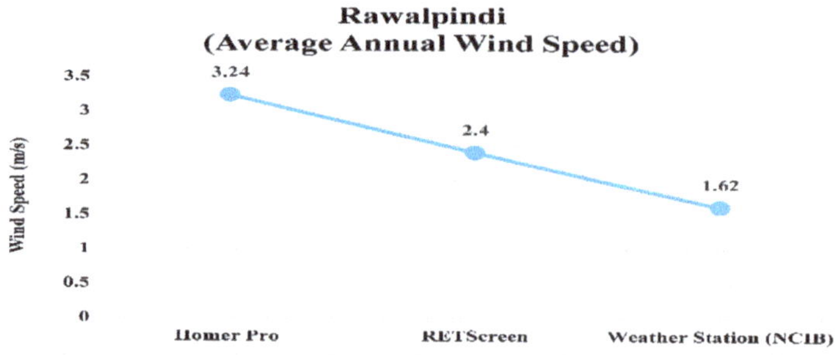

Figure 4. Comparison and validation of average annual wind speed data between HOMER Pro, RETScreen, and real-time data.

4. Conclusions

The maximum annual average solar irradiance of 5.3 & 5.06 (W/m^2) obtained from HOMER Pro and NREL, respectively, is found to be of Chakwal district. However, the maximum annual average wind speed potential of 3.48 & 2.5 (m/s) obtained from HOMER Pro and RETscreen , respectively, is of Jhelum district. The data of wind speed potential obtained from HOMER Pro, RETScreen, and Weather Station of NCIB is 3.24, 2.4, and 1.62, respectively. It can be concluded from the results that there is huge potential of solar irradiance in the districts of Potohar Plateau for the installation of PV system and its application in smart agricultural farms. However, wind potential is not sufficient for the installation of high-capacity and high-speed wind turbines.

Author Contributions: Conceptualization, M.S.K., S.H. and M.J.M.C.; methodology, M.S.K.; software, M.S.K. and T.I.; validation, M.S.K., S.H. and M.J.M.C.; formal analysis, M.S.K. and S.R.S.; investigation, S.H. and M.J.M.C.; resources, S.R.S. and M.S.W.; data curation, M.S.K., S.H., M.J.M.C. and T.I.; writing—original draft preparation, M.S.K. and S.H.; writing—review and editing, M.S.K., S.H., M.J.M.C., T.I., S.R.S. and M.S.W.; visualization, T.I., S.R.S. and M.S.W.; supervision, M.J.M.C., T.I., S.R.S. and M.S.W.; project administration, S.H., M.J.M.C., T.I., S.R.S.; funding acquisition, M.J.M.C. All authors have read and agreed to the published version of the manuscript.

Funding: This research received no external funding.

Institutional Review Board Statement: Not applicable.

Informed Consent Statement: Not applicable.

Data Availability Statement: Not applicable.

Acknowledgments: The authors acknowledge the National Center of Industrial Biotechnology (NCIB) for providing the weather data and funding support.

Conflicts of Interest: The authors declare no conflict of interest.

References

1. Ahmad, I. Chapter 14. In *2 Maccabees According to the Syriac Peshitta Version with English Translation*; Gorgias Press: Piscataway, NJ, USA, 2022; pp. 20–25. [CrossRef]
2. Lombardi, G.V.; Berni, R. Renewable energy in agriculture: Farmers willingness-to-pay for a photovoltaic electric farm tractor. *J. Clean. Prod.* **2021**, *313*, 127520. [CrossRef]
3. Gielen, D.; Boshell, F.; Saygin, D.; Bazilian, M.D.; Wagner, N.; Gorini, R. The role of renewable energy in the global energy transformation. *Energy Strateg. Rev.* **2019**, *24*, 38–50. [CrossRef]
4. Waseem Ghani, M.; Arshad, M.; Shabbir, A.; Shakoor, A.; Mehmood, N.; Ahmad, I. Investigation of potential water harvesting sites at potohar using modeling approach. *Pak. J. Agric. Sci.* **2013**, *50*, 723–729.
5. HOMER Pro–Microgrid Software for Designing Optimized Hybrid Microgrids. Available online: https://www.homerenergy.com/products/pro/index.html (accessed on 5 January 2022).
6. RETScreen. 2021. Available online: https://www.nrcan.gc.ca/maps-tools-and-publications/tools/modelling-tools/retscreen/7465 (accessed on 27 June 2022).

7. NASA. Available online: https://www.nasa.gov/ (accessed on 7 June 2022).
8. National Renewable Energy Laboratory (NREL) Home Page | NREL. Available online: https://www.nrel.gov/ (accessed on 7 June 2022).

environmental sciences proceedings

MDPI

Proceeding Paper

Sensing Techniques in Precision Agriculture for Pest and Disease Management †

Tayyaba Samreen [1,*], Aimen Tahir [1], Sidra-Tul-Muntaha [2], Muhammad Zulqernain Nazir [1], Muhammad Ahmad [1] and Sehrish Kanwal [3]

1 Institute of Soil and Environmental Sciences, University of Agriculture, Faisalabad 37000, Pakistan
2 Directorate of Pest Warning and Quality Control of Pesticides, Punjab, Lahore 54700, Pakistan
3 DDSDP (Data-Driven Smart Decision Platform), PMAS University of Arid Agriculture Rawalpindi, Rawalpindi 46000, Pakistan
* Correspondence: tybasamreen@gmail.com or tayyaba.samreen@uaf.edu.pk
† Presented at the 1st International Precision Agriculture Pakistan Conference 2022 (PAPC 2022)—Change the Culture of Agriculture, Rawalpindi, Pakistan, 22–24 September 2022.

Abstract: Precision agriculture (PA) is a cutting-edge, comprehensive, and globally recognized method. PA entails the application of agronomic ideas and modern technologies to improve sustainability, agricultural output, and environmental quality. This article is mostly concerned with sensing techniques. Remote sensing is a useful technique in PA for spotting, predicting, and predetermining the levels of infestations, and for controlling pests and diseases on a variety of fruits and crops. Sensors also transform traditional farming methods into precision farming methods, which helps to cut back on unnecessary input costs and raises agricultural productivity. By using these techniques, the application of pesticides can be quickly and locally administered.

Keywords: precision agriculture; remote sensing; sensors; pest management

Citation: Samreen, T.; Tahir, A.; Sidra-Tul-Muntaha; Nazir, M.Z.; Ahmad, M.; Kanwal, S. Sensing Techniques in Precision Agriculture for Pest and Disease Management. *Environ. Sci. Proc.* **2022**, *23*, 16. https://doi.org/10.3390/environsciproc2022023016

Academic Editor: Shoaib Rashid Saleem

Published: 20 December 2022

1. Introduction

Precision agriculture (PA) technologies have been developed and used over the past four decades, opening up new possibilities for crop and animal management by utilizing contemporary information technology. The current PA research also focuses on the creation of sensors that enable remote real-time crop and soil property detection, which includes remote sensing, digital image analysis, and biosensors. Therefore, a broader concept of PA encompasses all agricultural activities that use modern information technologies, including animal and plant production, precision livestock agriculture (PLA), soil fertility and water quality management, and raw material processing after harvest. PA technology also helps farmers control plant diseases and pests [1,2]. According to Zhang and his coworkers, PA is a creative, comprehensive, and globally standardized strategy for sustainable agriculture that seeks to improve resource use efficiency while reducing risks and management decision ambiguity [3]. It is a tactic that simultaneously addresses the requirements of both people and animals in terms of safety and well-being [4]. Precision agriculture applies artificial intelligence (AI), wireless communication, and information technologies, e.g., the Internet of Things (IoT), to agriculture for precisely managing the fertilization, plant pests in the fields, crop diseases, and irrigation [5]. The health monitoring of crops determines the farm's state regarding plant pests and is seen to be the primary application of smart agriculture [6,7].

2. Remote Sensing (RS) Systems

The use of remote sensing is now crucial for managing agriculture sustainably. Fitzgerald found that color shifts and subsequent changes in the canopy's appearance may be utilized to identify early mite infection in cotton fields using multispectral remote sensing.

For the IPM context in many agricultural techniques, remote sensing offers useful information. For instance, several airborne remote sensing techniques, such as multispectral and hyperspectral sensors, were employed for agricultural applications. The availability of imaging is made possible by the employment of either single-band cameras or video with appropriate filter lenses in these systems. Numerous writers looked at the use of airborne remote sensing for the analysis and detection of insect pest damage in the field. Additionally, it has been used to map agricultural diseases and evaluate their effects on crop yield [8,9].

3. Sensors

With the use of intelligent sensors, precision farming can increase agricultural output by giving farmers precise information and assisting them in making better decisions. Sensors involved in PA include green seekers or the Normalized Difference in Vegetative Index (NDVI) sensors, nano-biosensors, atomic dots (QDs), tiny barcodes, etc. [10] (Figure 1). Sensing technologies make it possible to record details on crop symptoms, ambient circumstances, or the physical traits of an insect pest or disease. Sensing techniques record ambient variables and pictures. Field sensors, spectroradiometers, and microscopes are the three different types of sensors employed to record the data in image form [11]. The next section explains each image recording category of sensors.

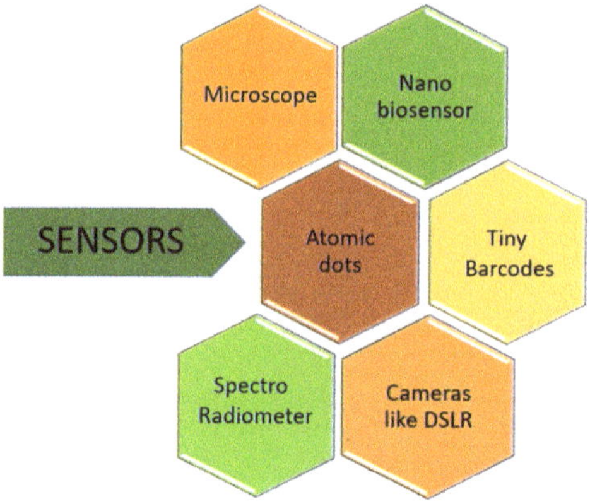

Figure 1. Different types of sensors.

- Field sensors: Ambient temperature sensors, image capture, soil-moisture sensors, humidity sensors, and water sensors are used to identify and manage diseases. The soil is also observed using environmental humidity, temperature, and moisture sensors. The findings demonstrated that the crop output is increased with prompt diagnosis and ongoing observation by using advanced techniques.
- Spectroradiometer: An experiment was carried out from 70 to 90 days after seeding that employed a spectroradiometer to identify and quantify the harm caused by T. tobacco (Lind). SVIs were calculated based on the reported canopy reflectance. Remote sensors were used to assess the relationship between a whitefly infestation and biotic stress. In the spectral range of 350–2500 nm at 1 nm, they employed a spectroradiometer with various sample intervals. To ascertain the correlation between the degree of damage caused by the whitefly infestation and chlorophyll content, the chlorophyll concentration was evaluated.
- Microscope: In natural settings, Rothe and Rothe utilized a DSLR camera and a Leica Wild M3C microscope to identify the presence of Myrothecium, Alternaria, and bacterial leaf blight [11].

4. Conclusions

The goal of PA research is to eliminate decision uncertainty brought on by unrestrained spatial and temporal variation. The secret to the successful future of PA, however, is not by simply gathering pertinent data, but also turning these data into knowledge that can be used to make choices and thoroughly weigh the risks and rewards of those actions. In the domain of sensors, cameras and outdoor sensors such as temperature and humidity sensors are the most popular types of sensors. Applications for remote sensing in agriculture are expanding and evolving, particularly in the management of pests and plant diseases. Remote sensing data are now viewed as general-purpose data for a large user population with potentially varied needs.

5. Future Outlooks

- Create prediction algorithms to determine the times and locations of disease and insect invasions.
- Prescriptive models should be used to specify methods to manage illness and insect pests.
- Create a system of sophisticated traps made of pheromones to anticipate the invasion of pests.
- Create illness diagnosis models.
- Create models with many detections for disease or insect assaults.

Author Contributions: T.S.: conceptualization, guiding, and improving the article. A.T. and S.-T.-M.: writing the original draft. M.Z.N., M.A. and S.K.: conceptualization, drafting, and revising the article. All authors have read and agreed to the published version of the manuscript.

Funding: This research received no external funding.

Institutional Review Board Statement: Not applicable.

Informed Consent Statement: Not applicable.

Data Availability Statement: No datasets were generated or analyzed for this article.

Acknowledgments: The authors are thankful to the 1st International Precision Agriculture Pakistan Conference held on 22–24 September 2022.

Conflicts of Interest: The authors declare no conflict of interest.

References

1. Karimzadeh, R.; Sciarretta, A. Spatial patchiness and association of pests and natural enemies in agro-ecosystems and their application in precision pest management: A review. *Precis. Agric.* **2022**, *23*, 1836–1855. [CrossRef]
2. Bae, S.; Yu, J.; Wang, L.; Kim, J.; Park, C. Experiments on unmanned aerial vehicle survey for detection of micro beach features. *J. Coast. Res.* **2019**, *90*, 354–361. [CrossRef]
3. Zhang, P.; Guo, Z.; Ullah, S.; Melagraki, G.; Afantitis, A.; Lynch, I. Nanotechnology and artificial intelligence to enable sustainable and precision agriculture. *Nat. Plants* **2021**, *7*, 864–876. [CrossRef] [PubMed]
4. Karar, M.E.; Alsunaydi, F.; Albusaymi, S.; Alotaibi, S. A new mobile application of agricultural pest's recognition using deep learning in cloud computing system. *Alex. Eng. J.* **2021**, *60*, 4423–4432. [CrossRef]
5. Hu, Z.; Xu, L.; Cao, L.; Liu, S.; Luo, Z.; Wang, J.; Li, X.; Wang, L. Application of Non-Orthogonal Multiple Access in Wireless Sensor Networks for Smart Agriculture. *IEEE Access* **2019**, *7*, 87582–87592. [CrossRef]
6. Karar, M.E.; al-Rasheed, M.; Al-Rasheed, A.; Reyad, O. IoT and Neural Network-Based Water Pumping Control System For Smart Irrigation. *Inform. Sci. Lett.* **2020**, *9*, 107–112.
7. Ayaz, M.; Ammad-Uddin, M.; Sharif, Z.; Mansour, A.; Aggoune, E.M. Internet-of-Things (IoT)-Based Smart Agriculture: Toward Making the Fields Talk. *IEEE Access* **2019**, *7*, 129551–129583. [CrossRef]
8. Dhau, I.; Adam, E.; Mutanga, O.; Ayisi, K.K. Detecting the severity of maize streak virus infestations in maize crop using in situ hyperspectral data. *Trans. R. Soc. S. Afr.* **2018**, *73*, 8–15. [CrossRef]
9. Dhau, I.; Adam, E.; Mutanga, O.; Ayisi, K.K.; Mutanga, O. Detection and mapping of maize streak virus using RapidEye satellite imagery. *Geocarto Int.* **2018**, *34*, 856–866. [CrossRef]

10. El-Ghany, A.; Nesreen, M.; El-Aziz, A.; Shadia, E.; Marei, S.S. A review: Application of remote sensing as a promising strategy for insect pests and diseases management. *Environ. Sci. Pollut. Res.* **2020**, *27*, 33503–33515. [CrossRef] [PubMed]
11. Rothe, P.; Rothe, J. Intelligent pattern recognition system with application to cotton leaf disease identification. *Innov. Comput. Sci. Eng.* **2019**, *32*, 19–27.

environmental sciences proceedings

MDPI

Proceeding Paper

Role of Nanotechnology in Precision Agriculture †

Tayyaba Samreen [1,*], Sehar Rasool [1], Sehrish Kanwal [2], Safia Riaz [3], Sidra-Tul-Muntaha [4] and Muhammad Zulqernain Nazir [1]

1 Institute of Soil and Environmental Sciences, University of Agriculture Faisalabad, Faisalabad 37000, Pakistan
2 DDSDP (Data-Driven Smart Decision Platform), PMAS University of Arid Agriculture Rawalpindi, Rawalpindi 46000, Pakistan
3 Pesticide Quality Control Lab, Ayub Agricultural Research Institute, Faisalabad 37000, Pakistan
4 Directorate of Pest Warning and Quality Control of Pesticides, Lahore 54700, Pakistan
* Correspondence: tybasamreen@gmail.com or tayyaba.samreen@uaf.edu.pk
† Presented at the 1st International Precision Agriculture Pakistan Conference 2022 (PAPC 2022)—Change the Culture of Agriculture, Rawalpindi, Pakistan, 22–24 September 2022.

Abstract: Nanotechnology is an interdisciplinary study field that attempts to boost agricultural output through substantial nanotechnology. This study has been conducted because of the reckless use of pesticides and synthetic fertilizers brought on by the green revolution, which has diminished soil biodiversity and increased disease and insect resistance. Only nanoparticles or nano chips can produce sophisticated biosensors for precision farming and deliver ingredients to plants in a nanoparticle-mediated manner. The precise distribution of nutrients and agrochemicals to plants is made possible by nano-encapsulated versions of conventional fertilizers, insecticides, and herbicides. Nanotechnology-based tests for detecting plant viral diseases are also gaining popularity and are useful for making a rapid and accurate diagnosis of viral disorders. The advantages and future uses of nanotechnology in precision agriculture are covered in this article. Modern technologies and methods based on nanotechnology can solve many issues in traditional agriculture and could revolutionize this industry.

Keywords: precision agriculture; nanotechnology; nanoparticles; nanomaterial

Citation: Samreen, T.; Rasool, S.; Kanwal, S.; Riaz, S.; Sidra-Tul-Muntaha; Nazir, M.Z. Role of Nanotechnology in Precision Agriculture. *Environ. Sci. Proc.* **2022**, 23, 17. https://doi.org/10.3390/environsciproc2022023017

Academic Editor: Muhammad Naveed Tahir

Published: 20 December 2022

1. Introduction

Precision agriculture (PA), which has several potential advantages in productivity, crop quality, profitability, sustainability, environmental protection, food safety, on-farm quality of life, and rural economic development, is revolutionizing agriculture all over the world. PA has become a crucial part of the strategy to accomplish this aim [1]. Nanotechnology is already widely used in modern agriculture, and precision agriculture is now a reality. Nanotechnology includes particles with dimensions of 100 nm or lower. Nanomaterials are used in the protection of plants, and in feeding and farming technique management because of their small sizes, high surface areas, and distinctive optical properties [2]. Plants react differentially to the nanoparticles regarding growth and metabolic activities. Nanoencapsulation is crucial for environmental protection because it stops hazardous substances from evaporating and leaking into the environment. To solve this problem, more effective, non-persistent insecticides must be developed, such as controlled-release formulations [3].

2. Examples of Nanoparticles

The production of nanoparticles from several biological sources has been exploited in agriculture for precision farming [4]. Here are some examples of these nanoparticles:

- Silver nanoparticles: When compared with commercial silver, silver nanoparticles have a higher antibacterial impact because of their large surface area and atoms. A wide variety of human diseases have been targeted by silver nanoparticles' antimicrobial properties [5]. As a result, there is increased interest in using silver nanoparticles' antibacterial properties to manage plant diseases [6].

- Zinc oxide nanoparticles: In alkaline soils containing calcium carbonate, zinc insufficiency is a prevalent micronutrient issue, negatively affecting agricultural productivity. To remediate the zinc shortage in soils, zinc fertilizers, such as zinc sulfate and zinc oxides, are often utilized.
- Titanium dioxide nanoparticles (TiO_2): Titanium is a sturdy metal that resists corrosion [7]. Titanium increases the formation of more carbohydrates, which promotes plant development and increases the rate of photosynthesis [8].

3. Nanotechnology and Its Role

- Fertilizer delivery: The massive volumes of fertilizers delivered have greatly enhanced yield but they also harm healthy soil microbiota. Run-off and pollution make fertilizers unavailable to plants. This issue can be resolved with nanomaterial-coated fertilizers (Figure 1).

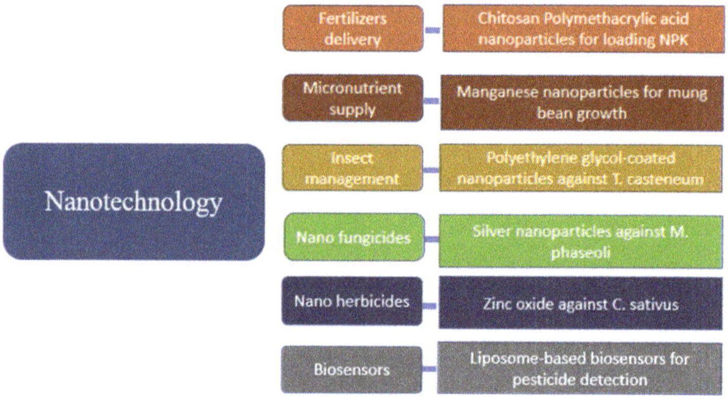

Figure 1. Role of nanotechnology.

- Micronutrient supply: Micronutrients, including zinc, iron, manganese, boron, copper, molybdenum, etc., are known to be crucial for growth and development. The green revolution's significant increase in crop yields and new agricultural techniques have led to a steady decline in soil micronutrients, including zinc, iron, and molybdenum. Micronutrient treatment on the leaves can improve absorption. Nano formulations of micronutrients can be applied as a spray to plants or added to the soil for root absorption to improve soil quality and strength [9].
- Insect pest management: Although synthetic agrochemicals have transformed agriculture, they have also created a new problem, namely insect pest resistance. For the control and supervision of insect pests in contemporary agriculture, nanoparticles hold enormous potential. PEG-coated nanoparticles have improved the insecticidal efficacy of garlic oil against Tribolium castaneum (the red flour beetle).
- Nano fungicides: Crop fungus infections significantly reduce crop productivity. Although there are several commercially accessible fungicides, their usage harms plants as well. Nanotechnology has significant potential to help with this issue. Antifungal drugs made of nanoparticles have been tested against harmful fungi. Silver nanoparticles showed a stronger antifungal impact with lower concentrations than titanium dioxide and zinc oxide nanoparticles [10].
- Nano herbicides: The largest challenge to agriculture is weeds, which reduce agricultural productivity by consuming nutrients that would otherwise be available to crop plants. Nano herbicides can play an influential role in the environmentally

benign removal of weeds from crops. Environmental safety is also achieved by the encapsulation of herbicides in polymeric nanoparticles [11].

- Biosensors: Utilizing computers, sensors, remote sensing tools, global positioning systems, and precision farming to locate environmental variables to ascertain if crops are growing as efficiently as possible or accurately pinpoint the type and location of issues.

4. Conclusions

When it comes to agriculture and agri-tech, nanotechnology is the key that unlocks the next revolutionary step. It has the potential to improve agricultural output while simultaneously fostering a more harmonious connection between chemical use and environmental equilibrium. Therefore, it has the potential to be employed as a state-of-the-art technology to assist in solving global hunger issues in the not-too-distant future. Several scientists believe that high-tech farms equipped with nanotechnology's smart nanotools will allow for greater productivity with fewer resources invested. It encourages the development of novel, effective agrochemicals for plants, such as nanofertilizers and nanopesticides, which help sustainably smart agriculture by inhibiting plant diseases and preventing crop failure.

5. Future Outlooks

- The use of nanotechnologies in agriculture could make a significant contribution by addressing the issue of sustainability and climate change;
- In reality, the use of nano-scale transporters and chemicals can improve the efficient use of pesticides and fertilizers, lowering the quantity that must be sprayed while maintaining yield;
- Nanosensor technologies can promote the spread of precision agriculture and have an influence on waste reduction, leading to both the reuse of waste and more production;
- To obtain consumer approval and support for this technology, it will be crucial to involve all stakeholders, consumer groups, and non-government sectors in an open discussion.

Author Contributions: T.S.: conceptualization, guiding and improving the article. S.R. (Sehar Rasool) and S.K.; writing the original draft. S.R. (Safia Riaz), S.-T.-M. and M.Z.N.: conceptualization and drafting and revising the article. All authors have read and agreed to the published version of the manuscript.

Funding: This research received no external funding.

Institutional Review Board Statement: Not applicable.

Informed Consent Statement: Not applicable.

Data Availability Statement: No datasets were generated or analyzed for this article.

Acknowledgments: The authors are thankful to the 1st international precision agriculture Pakistan conference held on 22–24 September 2022.

Conflicts of Interest: The authors declare no conflict of interest.

References

1. Zhang, P.; Guo, Z.; Ullah, S.; Melagraki, G.; Afantitis, A.; Lynch, I. Nanotechnology and artificial intelligence to enable sustainable and precision agriculture. *Nat. Plants* **2021**, *7*, 864–876. [CrossRef] [PubMed]
2. Ghormade, V.; Deshpande, M.V.; Paknikar, K.M. Perspectives for nano-biotechnology enabled protection and nutrition of plants. *Biotechnol. Adv.* **2011**, *29*, 792–803. [CrossRef] [PubMed]
3. Ragaei, M.; Sabry, A.H. Nanotechnology for insect pest control. *Int. J. Sci. Environ. Technol.* **2014**, *3*, 528–545.
4. Bansal, P.; Duhan, J.S.; Gahlawat, S.K. Biogenesis of nanoparticles: A review. *Afr. J. Biotechnol.* **2014**, *13*, 2778–2785.
5. Oves, M.; Khan, M.S.; Zaidi, A.; Ahmed, A.S.; Ahmed, F.; Ahmad, E.; Sherwani, A.; Owais, M.; Azam, A. Antibacterial and cytotoxic efficacy of extracellular silver nanoparticles bio fabricated from chromium reducing novel OS4 strain of *Stenotrophomonas maltophilia*. *PLoS ONE* **2013**, *8*, e59140. [CrossRef] [PubMed]

6. Mishra, S.; Singh, B.R.; Singh, A.; Keswani, C.; Naqvi, A.H.; Singh, H.B. Bio fabricated silver nanoparticles act as a strong fungicide against *Bipolaris sorokiniana* causing spot blotch disease in wheat. *PLoS ONE* **2014**, *9*, e97881. [CrossRef] [PubMed]
7. Sang, L.; Zhao, Y.; Burda, C. TiO_2 nanoparticles as functional building blocks. *Chem. Rev.* **2014**, *114*, 9283–9318. [CrossRef] [PubMed]
8. Chen, H.; Seiber, J.N.; Hotze, M. ACS select on nanotechnology in food and agriculture: A perspective on implications and applications. *J. Agric. Food Chem.* **2014**, *62*, 1209–1212. [CrossRef] [PubMed]
9. Peteu, S.F.; Oancea, F.; Sicuia, O.A.; Constantinescu, F.; Dinu, S. Responsive polymers for crop protection. *J. Polym. Sci.* **2010**, *2*, 229–251. [CrossRef]
10. Shyla, K.K.; Natarajan, N.; Nakkeeran, S. Antifungal activity of zinc oxide, silver and titanium dioxide nanoparticles against *Macrophomina phaseolina*. *J. Mycol. Plant Pathol.* **2014**, *44*, 268–273.
11. Kumar, S.; Bhanjana, G.; Sharma, A.; Sarita; Sidhu, M.C.; Dilbaghi, N. Herbicide loaded carboxymethyl cellulose nanocapsules as potential carrier in agrinanotechnology. *Sci. Adv. Mater.* **2015**, *7*, 1143–1148. [CrossRef]

environmental sciences
proceedings

MDPI

Proceeding Paper

Spatio-Temporal Assessment of Satellite-Based Precipitation Products for Hydroclimatic Applications over Potohar Region, Pakistan †

Muhammad Umer Nadeem [1,2,3], Muhammad Naveed Anjum [1,2,*], Muhammad Asif [3], Tahir Iqbal [2], Saddam Hussain [4,5], Hafiz Rana Azeem Sarwar [3] and Akhtar Abbas [3]

1 Department of Land and Water Conservation Engineering, Faculty of Agricultural Engineering and Technology, PMAS-Arid Agriculture University, Rawalpindi 46000, Pakistan
2 Data Driven Smart Decision Platform, PMAS-Arid Agriculture University, Rawalpindi 46000, Pakistan
3 Climate, Energy and Water Resources Institute, National Agriculture Research Center, Islamabad 44000, Pakistan
4 Department of Irrigation and Drainage, University of Agriculture Faisalabad, Faisalabad 38000, Pakistan
5 Department of Agricultural and Biological Engineering, University of California (UC Davis), Davis, CA 95616, USA
* Correspondence: naveedwre@uaar.edu.pk

† Presented at the 1st International Precision Agriculture Pakistan Conference 2022 (PAPC 2022)—Change the Culture of Agriculture, Rawalpindi, Pakistan, 22–24 September 2022.

Citation: Nadeem, M.U.; Anjum, M.N.; Asif, M.; Iqbal, T.; Hussain, S.; Sarwar, H.R.A.; Abbas, A. Spatio-Temporal Assessment of Satellite-Based Precipitation Products for Hydroclimatic Applications over Potohar Region, Pakistan. *Environ. Sci. Proc.* **2022**, *23*, 18. https://doi.org/10.3390/environsciproc2022023018

Academic Editor: M. Jehanzeb Masud Cheema

Published: 26 December 2022

Abstract: Reliable precipitation data at appropriate resolutions are essential for crop irrigation scheduling, climate change assessments, and mitigating floods and droughts. However, information on precipitation at acceptable spatial and temporal scales is lacking in Pakistan, particularly in undulating areas such as the Potohar region. Therefore, an investigation was conducted to check the accuracies of two satellite-based precipitation products (SPPs), PERSIANN-CDR and TRMM, concerning the gauge-based data obtained from the Pakistan Metrological Department (PMD). At the pixel and regional scales, the uncertainty in satellite-based daily, monthly, and annual precipitation estimates was investigated. Statistical indices were used for the spatio-temporal assessment of SPPs. The results reveal that both SPPs performed better on the monthly and daily scales. The TRMM precipitation product was in better agreement with in situ gauge-based data (with CC > 0.7). Therefore, we recommend using TRMM data for hydro-climatic applications in the Potohar region of Pakistan.

Keywords: satellite-based precipitation; PERSIAN-CDR; TRMM; Potohar region

1. Introduction

Monitoring precipitation data are crucial for making decisions regarding agricultural irrigation, the design of hydrological structures, and the provision of drinking water to urban and rural areas. Precipitation is considered a significant source of water used in rainfed agriculture. It is also one of the key climatic factors in Pakistan's socio-economic development, hydropower generation, development of hydrological structures (small/mini dams), rainwater harvesting, flood monitoring, and many other essential activities [1]. Pakistan's economy is said to be based on agriculture, which heavily relies on water. A little over 40% of the annual budget is allocated to agriculture, although dryland farming still receives little attention. Approximately 95% of stored water is used in many agricultural activities [2]. It is also a plentiful resource for the nation's health and growth, which has been under massive strain due to the fast-growing population over the past 60 years [2–4]. Therefore, it is essential to explore different sources of rainfall data. Radars and in situ metrological stations are commonly considered the most trusted sources for many applications [5], and Pakistan Metrological Department provides gauge information. According to the literature, one gauging station can accurately cover an area of about 36 km^2 [5,6]. A similar gauge pattern was observed in the Potohar region, indicating the poor presentation of data. Therefore, we must explore different sources of precipitation data.

Rainfed agriculture depends on rainwater and its conservation, so this study is more crucial in the rainfed Potohar area. The accuracy of SPPs depends on the local terrain and meteorological system, although they can provide continuous information about global precipitation at precise spatial and temporal resolutions [5–7]. Therefore, it is essential to assess the reliability of SPPs before directly applying them to various hydrological applications. Most contemporary SPPs use extensive precipitation estimation techniques to predict precipitation estimates. These techniques can offer reliable data at fine spatio-temporal scales, employing signals from infrared (IR) and microwave (MW) sensors [8–11].

In recent years, many SPPs have been evaluated in the mountainous ranges of Pakistan [2–5]. However, the spatial and temporal analysis of the two latest SPPs, including PERSIANN-CDR and TRMM, in the Potohar region of Pakistan, has not yet been evaluated. Therefore, the primary goal of this analysis is to assess satellite precipitation products (PERSIANN-CDR and TRMM) over the Potohar region of Pakistan. The issue of discontinuous precipitation data is anticipated to be solved by SPPs. Two recent SPPs (TRMM and CDR) over Pakistan's Potohar region will be assessed for the first time in terms of spatial and temporal resolutions. The conclusions of this research will be highly beneficial to Pakistani policymakers, hydrologists, meteorologists, water conservation measures, and users of SPPs data.

2. Materials and Methods

The Potohar region is situated between the Indus River and the Jhelum River at 33°30′0″ N and 73°0′0″ E and stretches from the salt range northward to the foothills of the Himalayas. Four districts—Attock, Jhelum, and Rawalpindi, and Chakwal— make up the Potohar region. Figure 1 depicts the study area's Digital Elevation Model (DEM) and available meteorological stations.

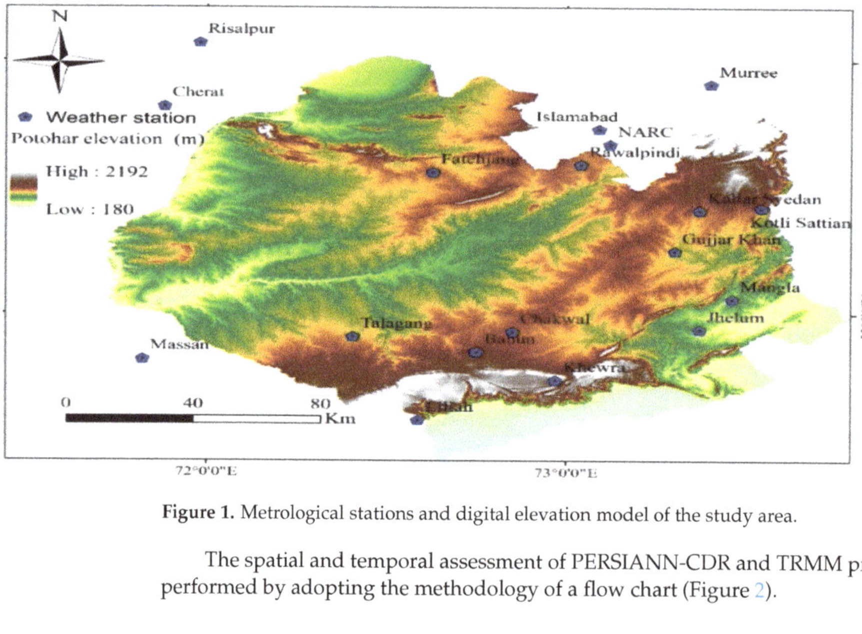

Figure 1. Metrological stations and digital elevation model of the study area.

The spatial and temporal assessment of PERSIANN-CDR and TRMM precipitation is performed by adopting the methodology of a flow chart (Figure 2).

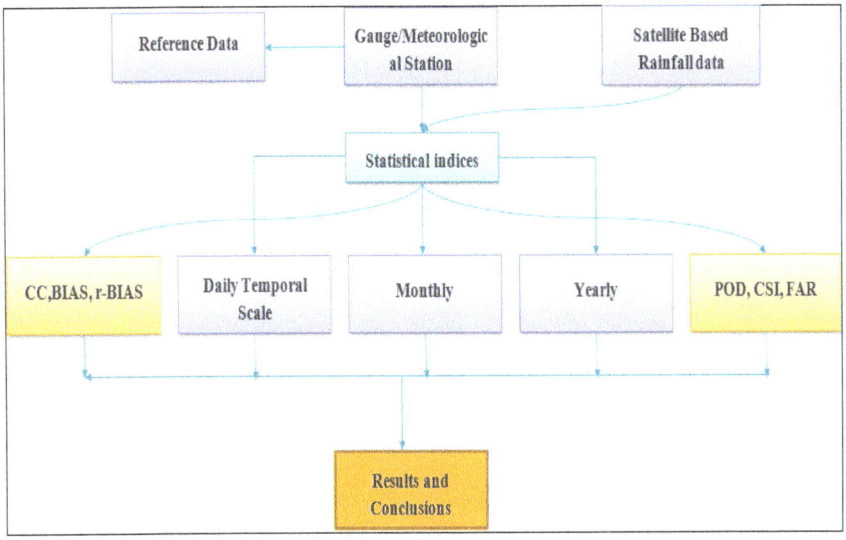

Figure 2. Layout of study.

3. Results and Conclusions

The spatial-temporal assessment analysis of PERSIANN-CDR and TRMM was conducted at a point-to-pixel scale over the Potohar region [5,6]. All SPPs were evaluated between 2017 and 2020. The main results of our study are illustrated in Figure 3.

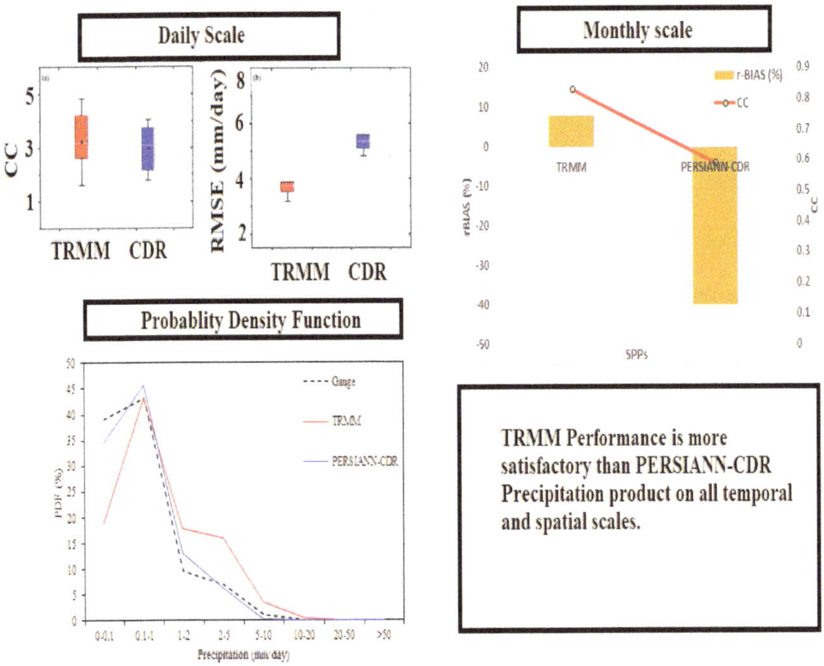

Figure 3. Results of TRMM and PERSIANN-CDR on different temporal and spatial scales.

- On a daily scale, the performance of TRMM is more satisfactory than PERSIANN-CDR.
- On a monthly scale, the precipitation estimates of both SPPs (TRMM and PERSIANN-CDR) are in good contact with gauge data.
- The daily PERSIANN-CDR performance is unsatisfactory (CC < 0.7). However, its monthly CC value (CC = 0.83) is acceptable.
- The error value RMSE is also the maximum for PERSIANN-CDR compared to TRMM.
- The TRMM outperforms the PERSIANN product based on a yearly scale.
- The POD and FAR are close to 1 for TRMM, but the PERSIANN-CDR values are not in acceptable ranges.
- PERSIANN-CDR has a higher false alarm ratio (FAR) than TRMM.

On daily, monthly, and annual scales, the CC values of the TRMM product are greater than 0.7, and its rBIAS and RMSE are in acceptable ranges. Therefore, we recommend using the TRMM product for hydroclimate applications over the Potohar region of Pakistan.

Author Contributions: All authors equally contributed to this article. All authors have read and agreed to the published version of the manuscript.

Funding: This study receive no external funding.

Institutional Review Board Statement: Not applicable.

Informed Consent Statement: Not applicable.

Data Availability Statement: Not applicable.

Conflicts of Interest: The authors declare no conflict of interest.

References

1. Anjum, M.N.; Irfan, M.; Waseem, M.; Leta, M.K.; Niazi, U.M.; Rahman, S.; Ghanim, A.; Mukhtar, M.A.; Nadeem, M.U. Assessment of PERSIANN-CCS, PERSIANN-CDR, SM2RAIN-ASCAT, and CHIRPS-2.0 Rainfall Products over a Semi-Arid Subtropical Climatic Region. *Water* **2022**, *14*, 147. [CrossRef]
2. FAO (Food and Agriculture Organisation of the United Nations). AQUASTAT. 2021. Available online: https://www.fao.org/aquastat (accessed on 3 September 2021).
3. Ashraf, M. Water Scarcity in Pakistan: Issues and Options. 2018. Available online: http://pcrwr.gov.pk/wp-content/uploads/2021/07/Water-Scarcity-in-Pakistan-Issues-and-Options-May-18.pdf (accessed on 16 August 2021).
4. Nadeem, M.U.; Waheed, Z.; Ghaffar, A.M.; Javaid, M.M.; Hamza, A.; Ayub, Z.; Nawaz, M.A.; Waseem, W.; Hameed, M.F.; Zeeshan, A.; et al. Application of HEC-HMS for flood forecasting in Hazara catchment Pakistan, south Asia. *Int. J. Hydrol.* **2022**, *6*, 7–12. [CrossRef]
5. Nadeem, M.U.; Anjum, M.N.; Afzal, A.; Azam, M.; Hussain, F.; Usman, M.; Javaid, M.M.; Mukhtar, M.A.; Majeed, F. Assessment of Multi-Satellite Precipitation Products over the Himalayan Mountains of Pakistan, South Asia. *Sustainability* **2022**, *14*, 8490. [CrossRef]
6. Nadeem, M.U.; Ghanim, A.A.J.; Anjum, M.N.; Shangguan, D.; Rasool, G.; Irfan, M.; Niazi, U.M.; Hassan, S. Multiscale Ground Validation of Satellite and Reanalysis Precipitation Products over Diverse Climatic and Topographic Conditions. *Remote Sens.* **2022**, *14*, 4680. [CrossRef]
7. Ali, A.F.; Xiao, C.; Anjum, M.N.; Adnan, M.; Nawaz, Z.; Ijaz, M.W.; Sajid, M.; Farid, H.U. Evaluation and comparison of TRMM multi-satellite precipitation products with reference to rain gauge observations in Hunza River basin, Karakoram Range, northern Pakistan. *Sustainability* **2017**, *9*, 1954. [CrossRef]
8. Nadeem, M.U.; Rasool, G.; Nawaz, M.A. Runoff Estimations by Using Satellite-Based Rainfall Products (Srps) and Then, Identifying Its Relation with Other Gauge Data Over Potohar Plateau, Pakistan. *Int. J. Water Res.* **2022**, *4*, 126. [CrossRef]
9. Amorim, J.d.S.; Viola, M.R.; Junqueira, R.; de Oliveira, V.A.; de Mello, C.R. Evaluation of satellite precipitation products for hydrological modeling in the Brazilian cerrado biome. *Water* **2020**, *12*, 2571. [CrossRef]
10. Tahir, A.A.; Chevallier, P.; Arnaud, Y.; Ahmad, B. Snow cover dynamics and hydrological regime of the Hunza River basin, Karakoram Range, Northern Pakistan. *Hydrol. Earth Syst. Sci.* **2011**, *15*, 2275–2290. [CrossRef]
11. Yang, M.; Li, Z.; Anjum, M.N.; Gao, Y. Performance evaluation of version 5 (V05) of Integrated Multi-satellite Retrievals for Global Precipitation Measurement (IMERG) over the Tianshan Mountains of China. *Water* **2019**, *11*, 1139. [CrossRef]

environmental sciences proceedings

MDPI

Proceeding Paper

Integration of Precision Agriculture Techniques for Pest Management †

Sehrish Kanwal [1,*], Muhammad Azam Khan [1,2], Shoaib Saleem [1], Muhammad Naveed Tahir [1], Sidra Tul Muntaha [3], Tayyaba Samreen [4], Sidra Javed [5], Muhammad Zulqernain Nazir [4] and Basit Shahzad [1,2]

1 Data-Driven Smart Decision Platform, PMAS-Arid Agriculture University, Rawalpindi 46000, Pakistan
2 Department of Horticulture, Faculty of Crop and Food Sciences, PMAS-University of Arid Agriculture, Rawalpindi 46000, Pakistan
3 Directorate of Pest Warning and Quality Control of Pesticides, Lahore 39100, Pakistan
4 Institute of Soil and Environmental Sciences, University of Agriculture, Faisalabad 38040, Pakistan
5 National Centre for Industrial Biotechnology (NCIB), PMAS-University of Arid Agriculture, Rawalpindi 46000, Pakistan
* Correspondence: sehrishkanwal7777@gmail.com
† Presented at the 1st International Precision Agriculture Pakistan Conference 2022 (PAPC 2022)—Change the Culture of Agriculture, Rawalpindi, Pakistan, 22–24 September 2022.

Abstract: Horticultural crops have a special impact on a nation's economy due to their significance in raising the living standards of farmers. The traditional method for crop protection is becoming ineffective with an increase in climate change effects. Precision Agriculture (PA) presents a solution to this issue through the precise monitoring and forecasting of pests, which improves productivity and guarantees environmental sustainability. The productivity of the existing plant production system can be increased by using precision agriculture techniques. Various PA technologies that enable farmers to monitor the pest include remote sensing, the Internet of Things, geographical information systems, and artificial intelligence. The PA technique assists with pest forecasting and the management of pests and diseases in plants. The content of this article was gathered through a literature review of recent research. The approaches used in PA for pest forecasting, monitoring, and management are the main topic of this study. In the long term, this study will help farmers to manage insect pests in a way that is both affordable and environmentally beneficial.

Keywords: precision agriculture; pest forecasting; pest monitoring; pest management

Citation: Kanwal, S.; Khan, M.A.; Saleem, S.; Tahir, M.N.; Muntaha, S.T.; Samreen, T.; Javed, S.; Nazir, M.Z.; Shahzad, B. Integration of Precision Agriculture Techniques for Pest Management. *Environ. Sci. Proc.* **2022**, *23*, 19. https://doi.org/10.3390/environsciproc2022023019

Academic Editor: Muhammad Naveed Anjum

Published: 26 December 2022

1. Introduction

The yield and quality of citrus and olive crops are significantly harmed by insect infestation. Different parts of the plant are inhabited by insect pests, hampering the productivity and survival of plants. Therefore, to prevent pest attacks and production losses, the pest management techniques need to be updated regularly to ensure better output and environmental sustainability [1]. Remote Sensing (RS), the Internet of Things (IoT), Geographic Information Systems (GIS), and Artificial Intelligence (AI) are just a few of the PA technologies that give farmers the ability to monitor pests. According to Moses-Gonzales, N and M.J. Brewer, [2], sensing drones along with sensors can detect and identify pest-caused damage, pest habitat, and pest presence. This study is being carried out to gather information about the main citrus and olive [3] orchard pests that exist worldwide. This study investigated the major PA tools available for pest management. Furthermore, this study reviewed the literature about monitoring and forecasting approaches available for citrus and olive pest monitoring and forecasting.

2. Methodology

A review of secondary published data was conducted to retrieve authentic information for this investigation. To gather relevant information about the pest management system, various research papers, books, web-based publications, and initiatives were reviewed. The study utilises recent data about the integration of PA in pest monitoring, forecasting, and management.

3. Role of Precision Techniques in Pest Management

Forecasting pest activity is made easier by precision farming techniques. According to automatic pest control systems, pests can be identified by using infrared sensors, thermal sensors, audio sensors, and image-based categorisation. To increase crop productivity and sustainability, sensors and software are used to interpret the data. High-resolution crop photos from satellite or aerial platforms (manned or unmanned) were processed to extract more information, i.e., employed for making pest control decisions.

4. Major Precision Approach for Monitoring Pests

For the effective production of the crop, pest surveillance and management are crucial as described in Figure 1. A pest monitoring program is thought to be crucial for identifying various pests on a large scale. Modern monitoring tools for pest monitoring include GPS, Global Navigation Satellite System (GNSS), and AI.

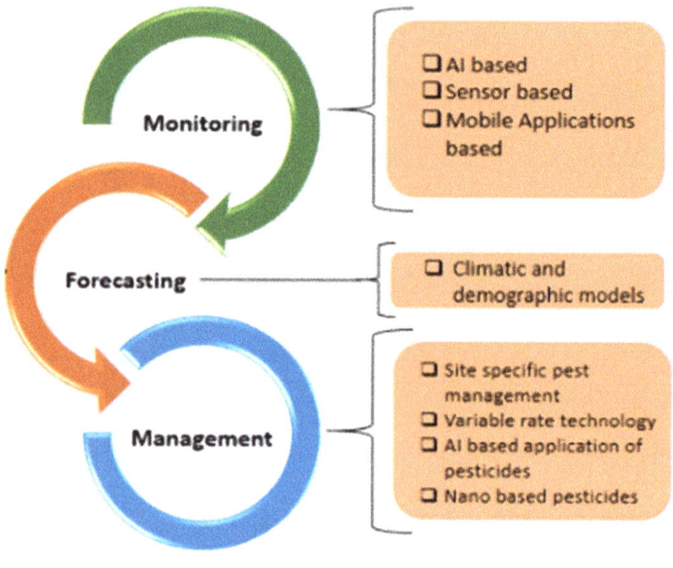

Figure 1. Precision Pest Management.

i. AI-based monitoring of pests: With its stringent skill set, AI has emerged as a crucial tool for addressing many issues in agriculture. Deep learning (DL) models for AI assist in identifying pests [4].
ii. Sensors-based monitoring of pests: The use of sensor-based PA technologies in field surveys of various pests has recently begun. Some of the examples include the use of radar to track the migration of pests, thermal infrared imaging, GNSS for telemetry of wildlife, video equipment to monitor flying insects, chemiluminescent tags to track insect movement at night, detection of larval movement through echo-sounding and habitat mapping.
iii. Mobile app for monitoring pests: Different smartphone applications are available for automated management and control of agricultural pests. The most popular are Trapview System, Plantix-Mobile App, Agri-App, and Kheti-Badi [5].

5. Role of Precision in the Forecasting of Pests

The prediction of pests helps farmers in reducing pesticide usage, avoiding blanket pesticide treatments, and providing high-quality outcomes by advising them on the biology and timing of insect occurrence [6]. The use of forecasts and early warnings based on biophysical techniques allows the management of near-pest assaults, which improves pest control, lowers crop loss, and lowers cultivation costs.

5.1. Climatic Model for Forecasting the Pest Attack

Climate influences both insect abundance and the number of natural enemies, which is a crucial organic component in pest population management. Climate models may be a useful tool for predicting the spectrum of potential changes on a global scale when paired with the environmental needs of a specific pest species.

5.2. Other Models for Forecasting the Attack

Pest forecasting mostly utilizes an embedded system and an intelligent system for plant protection and pest management [7]. The important techniques used in pest forecasting are:

1. The use of machine learning (ML) techniques enables users to handle complicated information and conduct trend analysis for a variety of applications (such as data mining, image processing, and predictive analytics). A machine learning (ML) model was used to forecast the appearance of the summer generation of pests.
2. Physiologically based demographic models (PBDMs) might be utilized effectively in the context of climate change. A PBDM model is utilized to observe the weather pattern, the location, and the abundance of pests.

When examining the insect population dynamics, the most important aspect is to consider both the local weather conditions and worldwide climate indicators. Moreover, the combined effects of exogenous (such as regional climate conditions) and endogenous (such as intrinsic population dynamics) components are also examined and modelled for pest forecasting [3].

6. Management of Insect Pests Using Precision Agriculture Tool

The goal of pest control in agricultural production is to keep pest populations or damage at levels that are both socially acceptable and economical. Some of the management techniques include:

1. Site-specific insect pest management: This technique involves the management of pests at a specific location, rather than treating the entire region. The advancement of AI helps us to manage pests at a specific location.
2. Variable rate technologies (VRT): Variable rate technologies allow farmers to apply pesticides at a specific location by utilizing AI and GIS techniques. This reduces the negative impact of pesticides on the environment and preserves beneficial insects.
3. AI-based application of pesticides: Artificial intelligence is the most recent technique employed to manage pests. The most common AI-based technologies used in pest management are unmanned aircraft systems. The first unmanned aerial vehicle (UAV) to be used for commercial purposes was developed by Yamaha Motor Co. Ltd. in Shizuoka, Japan, to monitor pest attacks in orchards.
4. Nano-bio-pesticides: Nanotechnology improves pest management through the effective utilization of pesticides. The size and dimensions of the nano pesticides allow the farmers to apply them precisely on crops.

7. Future Perspectives of PA for Insect-Pest Management

The broad literature review reveals the following research gaps in pest control that must be appropriately addressed if accurate findings are to be anticipated:

- A few prediction models for pests are not researched;

- The use of sensors for pest prediction has not yet been properly investigated;
- The information on weather-related pest occurrence in rain-fed crops is insufficient;
- Farmers do not receive the most recent forecast information timely.

8. Conclusions

The review shows that remote sensing (sensors), a precision tool, is used to assess agricultural variability, monitor crop health, and identify insect outbreaks. As early identification and correction to inadequate abiotic conditions may avoid significant pest outbreaks, they might act as decision support aids. Engineers, ecologists, and agronomists must collaborate on multidisciplinary research for this approach, which has huge market viability.

Author Contributions: Conceptualization, S.K., S.S. and M.A.K.; methodology, S.K. and M.Z.N.; investigation, S.S. and B.S.; literature discussion, S.S., M.N.T.; writing—original draft, S.K., S.T.M. and S.J.; writing—review and editing, T.S.; supervision, S.S.; project administration, M.N.T.; funding acquisition, S.K. All authors have read and agreed to the published version of the manuscript.

Funding: The study is a part of PSDP-funded project No. PSDP-332 "Data-Driven Smart Decision Platform to increase Agriculture productivity".

Institutional Review Board Statement: Not applicable.

Informed Consent Statement: Not applicable.

Data Availability Statement: Not applicable.

Acknowledgments: The authors would like to thank the DDSDP project.

Conflicts of Interest: The authors declare no conflict of interest.

References

1. Bolo, B.; Mpoeleng, D. Mapping of crop birds pest using GPS and GIS. *J. Agric. Inform.* **2019**, *10*, 12–20. [CrossRef]
2. Moses-Gonzales, N.; Brewer, M.J. A Special Collection: Drones to Improve Insect Pest Management. *J. Econ. Entomol.* **2021**, *114*, 1853–1856. [CrossRef] [PubMed]
3. Ordano, M. Olive fruit fly (*Bactrocera oleae*) population dynamics in the Eastern Mediterranean: Influence of exogenous uncertainty on a monophagous frugivorous Insect. *PLoS ONE* **2015**, *10*, e0127798. [CrossRef] [PubMed]
4. Brisco, B. Precision agriculture and the role of remote sensing: A review. *Can. J. Remote Sens.* **1998**, *24*, 315–327. [CrossRef]
5. Toscano-Miranda, R.; Toro, M.; Aguilar, J.; Caro, M.; Marulanda, A.; Trebilcok, A. Artificial-intelligence and sensing techniques for the management of insect pests and diseases in cotton: A systematic literature review. *J. Agric. Sci.* **2022**, *160*, 16–31. [CrossRef]
6. Magarey, R.D. The NCSU/APHIS Plant Pest Forecasting System (NAPPFAST). In *Pest Risk Modelling and Mapping for Invasive Alien Species*; CABI: Wallingford, UK, 2015; pp. 82–96. [CrossRef]
7. Gaikwad, S.V. An innovative IoT based system for precision farming. *Comput. Electron. Agric.* **2021**, *187*, 106–116. [CrossRef]

environmental sciences proceedings

MDPI

Proceeding Paper

Application of Satellite Remote Sensing Data to Manage Groundwater in Irrigated Canal Zone of Punjab †

Aamir Shakoor [1,*], Muhammad Ali [1], Zahid Mahmood Khan [1], Hafiz Umer Farid [1] and Muhammad Abubakar Aslam [2]

1 Department of Agricultural Engineering, Bahauddin Zakariya University, Multan 60800, Pakistan
2 National Center of Industrial Biotechnology (NCIB), PMAS Arid Agriculture University, Rawalpindi 46000, Pakistan
* Correspondence: aamirskr@bzu.edu.pk
† Presented at the 1st International Precision Agriculture Pakistan Conference 2022 (PAPC 2022)—Change the Culture of Agriculture, Rawalpindi, Pakistan, 22–24 September 2022.

Abstract: Irrigated agriculture is highly dependent on groundwater resources in Pakistan. Due to the fiber and food requirements, the reliance on groundwater has been increasing during the past two decades. This research work was conducted at the canal command area of the Dera Ghazi Khan (D.G Khan) canal for years 2017–2019. There were two products of the satellite i.e., AQUA (EOS PM) and TERRA (EOS AM), which were used in this research for determining groundwater demand in the area. For Rabi seasons 2017–2018 and 2018–2019, cropping areas of the major crops i.e., wheat and sugarcane, were 98,712.5 and 131,856.2 ha, and 100,568.7 and 132,743.7 ha, respectively. For the Kharif seasons of 2018 and 2019, cropping areas of major crops i.e., rice, cotton, and sugarcane, were 82,093.7, 74,150, and 98,712.5 ha, and 75,687.5, 79,275, and 132,743.7 ha, respectively.

Keywords: groundwater demand; NDVI; AQUA; TERRA; Dera Ghazi Khan

Citation: Shakoor, A.; Ali, M.; Khan, Z.M.; Farid, H.U.; Aslam, M.A. Application of Satellite Remote Sensing Data to Manage Groundwater in Irrigated Canal Zone of Punjab. *Environ. Sci. Proc.* **2022**, *23*, 20. https://doi.org/10.3390/environsciproc2022023020

Academic Editor: Saddam Hussain

Published: 26 December 2022

1. Introduction

In Pakistan, groundwater has been used for irrigating agriculture for a long time. Earlier, the abstraction of groundwater was made through open wells with the help of rope and buckets, reciprocating pumps, and hand pumps. In Pakistan, for irrigation for agriculture, groundwater is the second-largest source because of arid climatic conditions [1]. Due to the meeting of fiber and food requirements in Pakistan, the reliance on groundwater has been increasing during the past two decades [2–4].

In Pakistan, groundwater is a very important resource because its demand is increasing day by day for agricultural, domestic, and industrial purposes. Across the country, the contribution of groundwater for agriculture purposes is about 60%, for drinking it is 90%, and for industry it is 100% [5]. As most of the climatic zone of Pakistan is an arid climate zone where average annual rainfall is less than 20 mm and groundwater is limited in these areas [6]. In the Punjab province of Pakistan, the installation rate of private tubewells has been increasing at the rate of 60% from 1991–2000. In Pakistan, 1.20 million tubewells are installed that have a discharge rate of 0.015–0.56 m^3/s, having a depth of 30–85 m and a diameter of 15–30 cm out of this, and almost 86% of tube wells are installed in Punjab province. The groundwater has great importance for the farmers of tail ends of distributaries and watercourses because groundwater supplies are not enough there [7–9]. The advances in remote-sensing availability of images and image acquisition are helpful for geoscientists to prepare maps quickly and explore and evaluate the geo-characteristics of any area on the globe.

The investigation of groundwater demand is one of the main components of the management of these resources. Thus, the remote sensing data, one of the latest techniques, was used to determine the crop water requirements and groundwater abstraction need in the study area.

2. Materials and Methods

2.1. Data

This research work was conducted at the canal command area of Dera Ghazi Khan (D.G Khan) canal, located at latitude 29.731° to 29.862° N and longitude 70.314° to 70.487° E. Elevation level of the study area is 129 m (423 ft) above the sea level. The canal network of the D.G. Khan Irrigation zone is 2679.64 miles long, and the command area under cultivation is about 0.84 million ha. The Chashma and Taunsa barrages provide irrigation water to the DG Khan canal.

The hottest month in this area is June, which has a maximum average temperature of 42 °C, and the coldest month is January, which has minimum average temperature of 6 °C. The average annual rainfall of this area is about 236.3 mm/year, having arid climate and erratic rainfall pattern. There are two seasons of crops, Kharif and Rabi. The Kharif season started in winter and is also named after monsoon crops because these crops are cultivated in the monsoon season. Sowing of Kharif crops started in April and harvesting started in October/November. Crops of the Kharif season are cotton, rice, and sugarcane. The Rabi season is sowing in November/December and harvesting in April/May. Major crops of the Rabi season are wheat, mustard, sugarcane, and barley.

2.2. MODIS Satellite Data

The remote sensing data were collected from the source of earth data, and its components were satellites AQUA (MYD13Q1) and TERRA (MOD13Q1), having a resolution of 250 m. The temperature data for the study area were collected from TERRA (MOD11A2), and rainfall data were collected online from weatheronline.com/dera-ghazi-khan for the time duration of 2017–2019. The discharge data of the canal were collected daily from the Punjab Irrigation Department for the time duration of 2017–2019. The MODIS satellite data set were acquired from its two sensors (AQUA and TERRA) and used for further processing. The data downloaded from these products have a difference of 8 days, and each product provides data after every 15 days.

2.3. Selection of Tiles of the Study Area

The selection of tiles in MODIS was very important because, in this method, the required tiles were selected in the study area that they lie in. The tiles were viewed from the app Google Earth that created a grid shape in Google Earth software around the globe, which was seen in red lines covering the area.

The interface of MODIS tiles is shown in Figure 1. The study area lies in horizontal tile h24 and two vertical tiles, v5 and v6. Therefore, the complete study area of research was covered by h24, v5, h24, and v6 tiles. Images through the satellite were easy for the graphical study of an area because tiles cover large areas, and Punjab (Pakistan) was completely covered by 2 tiles.

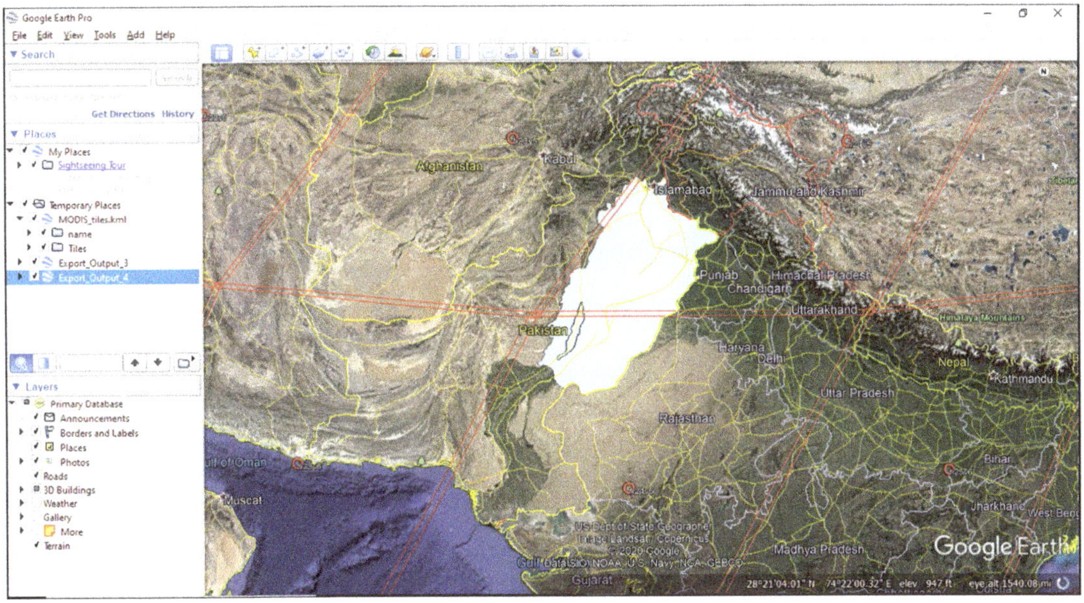

Figure 1. Selection of MODIS tiles of command area.

3. Results and Discussion

3.1. Seasonal Crop Water Requirement

The volume of seasonal crop water requirements for Rabi and Kharif is shown in Figure 2. The volume of crop water requirement ranged between 353.76 and 353.04 Mm^3 (million cubic meter) for the wheat crop during the season 2017–2018 and 2018–2019, respectively. The volume of crop water requirement ranged from 445.42 and 445.1 Mm^3 for the sugarcane crop for seasons 2018 and 2019, respectively. It was observed that the crop water requirement was high in 2017–2018 for Rabi and 2018 for Kharif, as compared to 2018–2019 and 2019, respectively.

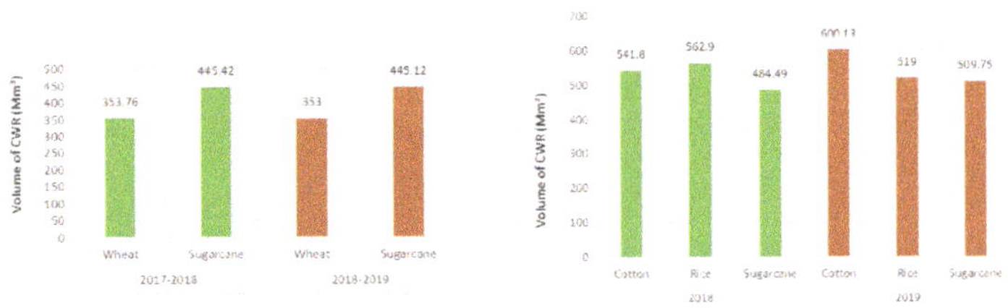

(**a**) (**b**)

Figure 2. Seasonal crop water requirements (CWR): (**a**) Rabi (**b**) Kharif.

3.2. Seasonal Groundwater Demand

The groundwater demand was calculated for each cropping season, as shown in Figure 3. The volume of groundwater demand for Rabi 2017–2018 and 2018–2019 was 642.24 and 661.57 Mm^3 and for Kharif 2018, 2019 was 165 and −101.18 Mm^3, respectively. The highest value of groundwater was required for season Rabi 2018–2019, and the lowest value for Kharif was in 2018. It was observed that Kharif 2019 showed a negative value in the graph, representing that there was no need for groundwater for this season and that canal water was enough to meet crop water requirements.

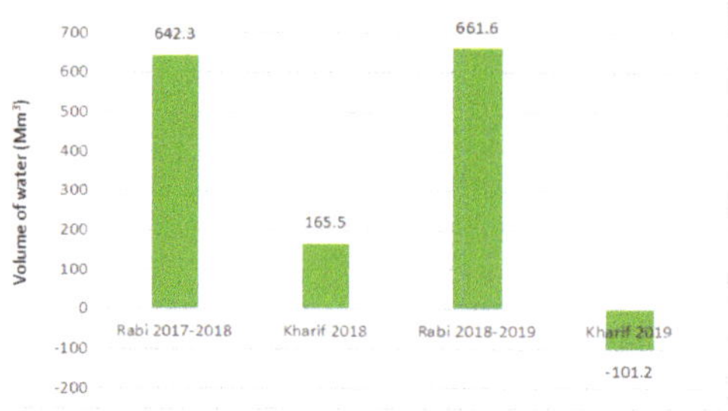

Figure 3. Seasonal groundwater demand.

4. Conclusions

In this research, AQUA and TERRA remotely sensed datasets were used to determine crop water requirements of the major crops in the study area for the year 2017–2019. The groundwater demand was calculated by subtracting the CWR from the actual canal water supplies. It was found that the seasonal volume of CWRs of wheat and sugarcane for season 2017–2018 were 353.76 and 445.42 Mm^3 and for season 2018–2019 were 353 and 445.12 Mm^3, respectively. The seasonal volume of CWRs for rice, cotton, and sugarcane for season 2018 were 562.9, 541.8, and 484.5 Mm^3, and for season 2019 they were 518.9, 600, and 509 Mm^3 respectively. The results showed that volume of groundwater demands for Rabi 2017–2018 and 2018–2019 were 642.24 and 661.57 and for Kharif 2018 was 165 Mm^3, respectively. For Kharif 2019, the negative values (-101.18 Mm^3) represent the surplus canal water supplies. The groundwater demand in the Rabi season was relatively higher than Kharif season. It is recommended that the maximum number of crops should be observed for a better understanding of the crop water requirements in the area.

Author Contributions: A.S. and M.A. conceptualize the research objectives. M.A., H.U.F. and M.A.A. collected the data. A.S. and Z.M.K. supervised the procedure and method adopted. All the authors contributed to analyzing the results and writing the manuscript. All authors have read and agreed to the published version of the manuscript.

Funding: This research received no external funding.

Institutional Review Board Statement: Not applicable.

Informed Consent Statement: Not applicable.

Data Availability Statement: The data are available upon request from the corresponding author.

Acknowledgments: The authors acknowledge the Department of Agricultural Engineering, Bahauddin Zakariya University, Multan for proving facilities to conduct this research.

Conflicts of Interest: The authors declare no conflict of interest.

References

1. Imran, M.A.; Ali, A.; Ashfaq, M.; Hassan, S.; Culas, R.; Ma, C. Impact of Climate Smart Agriculture (CSA) practices on cotton production and livelihood of farmers in Punjab, Pakistan. *Sustainability* **2018**, *10*, 2101. [CrossRef]
2. Ashfaq, A.; Ashraf, M.; Bahzad, A. Spatial and temporal assessment of groundwater behaviour in the Soan Basin of Pakistan. *Univ. Eng. Technol. Taxila Tech. J.* **2014**, *19*, 12.
3. Qureshi, A.S.; McCornick, P.G.; Sarwar, A.; Sharma, B.R. Challenges and prospects of sustainable groundwater management in the Indus Basin, Pakistan. *Water Resour. Manag.* **2010**, *24*, 1551–1569. [CrossRef]
4. Watto, M.A.; Mugera, A.W. Groundwater depletion in the Indus Plains of Pakistan: Imperatives, repercussions and management issues. *Int. J. River Basin Manag.* **2016**, *14*, 447–458. [CrossRef]

5. Hassan, G.Z.; Hassan, F.R. Sustainable use of groundwater for irrigated agriculture: A case study of Punjab, Pakistan. *Eur. Water* **2017**, *57*, 475–480.
6. Jayaram, D. India–Pakistan river water sharing: Prospects for environmental peacebuilding. In *Water Management in South Asia*; Springer: Cham, Switzerland, 2020; pp. 175–202.
7. Shakoor, A.; Mahmood Khan, Z.; Arshad, M.; Farid, H.U.; Sultan, M.; Azmat, M.; Hussain, Z. Regional groundwater quality management through hydrogeological modeling in LCC, West Faisalabad, Pakistan. *J. Chem.* **2017**, *2017*, 2041648. [CrossRef]
8. Nagrah, A.; Chaudhry, A.M.; Giordano, M. Collective action in decentralized irrigation systems: Evidence from Pakistan. *World Dev.* **2016**, *84*, 282–298. [CrossRef]
9. Awan, U.K.; Anwar, A.; Ahmad, W.; Hafeez, M. A methodology to estimate equity of canal water and groundwater use at different spatial and temporal scales: A geo-informatics approach. *Environ. Earth Sci.* **2016**, *75*, 409. [CrossRef]

environmental sciences proceedings

MDPI

Proceeding Paper

A Performance Comparison of Variable Rate Technologies for Spot-Specific and Uniform Spraying for Citrus Orchard †

Arshed Ahmed [1,*], Shahram Hamza [2], Muhammad Naveed Tahir [3], Shoaib Rashid Saleem [4] and Rashed Ahmed [1]

1 Department of Electronics, University of Peshawar, Peshawar 25000, Pakistan
2 Department of Electronics, University of Engineering and Technology Taxila, Attock 43600, Pakistan
3 Department of Agronomy, PMAS-Arid Agriculture University Rawalpindi, Rawalpindi 46000, Pakistan
4 Department of Farm Machinery and Precision Engineering, PMAS-Arid Agriculture University Rawalpindi, Rawalpindi 46000, Pakistan
* Correspondence: arshedahmed104@uop.edu.pk
† Presented at the 1st International Precision Agriculture Pakistan Conference 2022 (PAPC 2022)—Change the Culture of Agriculture, Rawalpindi, Pakistan, 22–24 September 2022.

Abstract: Shortly, the demand for food production will increase, and the challenges to agriculture, including climate change, land degradation, farmable land availability and labor effort, will be resolved by applying precision agriculture techniques. In modern precision agriculture, variable rate (VR) technology is one of the key research areas. This technology provides an improved adjustment of agrochemical dosage to the trees by applying pesticide spraying to only areas where trees are detected based on their heights. This paper presents the testing results of an indigenously developed spot-specific sprayer for orchards based on tree detection through ultrasonic sensors. Lab Results show that spot-specific spraying provides 40% of agrochemical savings for an area of 100 acres compared with traditional spraying for the citrus orchard. Such technology can facilitate the farmers by automatic pesticide sprayers and economically protect the plants.

Keywords: orchard sprayer; object detection; ultrasonic sensor; solenoid valves; micro-controller

Citation: Ahmed, A.; Hamza, S.; Tahir, M.N.; Saleem, S.R.; Ahmed, R. A Performance Comparison of Variable Rate Technologies for Spot-Specific and Uniform Spraying for Citrus Orchard. *Environ. Sci. Proc.* **2022**, *23*, 21. https://doi.org/10.3390/environsciproc2022023021

Academic Editor: Qamar Zaman

Published: 26 December 2022

1. Introduction

Today agricultural industries are adopting both primitive and modern sensors and controller-based technologies to tackle these issues, each having advantages and disadvantages. Modern technologies are also judged based on their impact on the global environment and ecosystem. In most parts of the world, uniform spraying of pesticides is practiced by the use of pesticide-filled tanks and nozzles for spraying. Mainly it is carried by a single person who sprays it across the field [1]. This method also has many disadvantages because uniform spraying does not consider plant size, and farmers rush mostly without proper protection, which may result in many health hazards. Such uniform spraying is primarily manual, which works on the hydraulics principle, which is exhausting. This spraying technique uses more pesticide than the required amount for the crop, which is expensive and time-consuming [2].

In the early 2000s, this manual spaying was replaced with electric pump spraying; it had reduced the human effort, but the rest was the same; for example, this electric pump spraying technique also used more pesticides than actually required for crops, which was the core problem [3].

Researchers have attempted to develop variable rate (VR) technologies for various crops. To date, less attention has been given to the economical use of ultrasonic sensors in variable rate (VR) systems. S. Gangadharan et al. conducted a comparative analysis between a LIDAR sensor and an ultrasonic sensor [4]. Sensors were tested in such a way that enabled the process of an automated shaker system and controlled its working. Both sensors were tested in an outdoor and indoor environment with accuracy factors of 10.16 and 5.08, respectively. The sensor was mounted in vertical and horizontal setups to develop a 3D canopy profile for a citrus tree groove. They also worked on volume estimation with the LIDAR sensor in their study. However, very little work has been performed to incorporate an ultrasonic sensor for tree detection in the intelligent sprayer [5].

In this paper, using an ultrasonic sensor integrated into a local boom sprayer that avoids unnecessary usage of expensive chemicals is presented that will not only reduce the cost of production but reduce environmental pollution. The boom sprayer provides better penetration of agrochemicals without damaging leaves and thus protects from various insects/diseases.

2. Methodology

The system is powered by a 12 V battery connected to a solar panel for automatic charging from sunlight. The system consists of 4 modules that are centrally controlled by a microcontroller. After boom selection, the sensors will start detecting objects at a certain distance, which can be altered according to a user's requirements.

The ultrasonic sensor will send a high flag signal to the microcontroller if a tree or object is detected within the sensing range of the sensor. The controller in the system will decide which relay to operate attached to that particular sensor's output. The relay will switch ON the solenoid valves connected to that specific relay, as shown in Figure 1b. The system is mounted on a metal frame boom 3 feet above the ground. The distance between each sensor is 2.5 feet, and the distance between the sensor and solenoid value is 10 inches during lab testing.

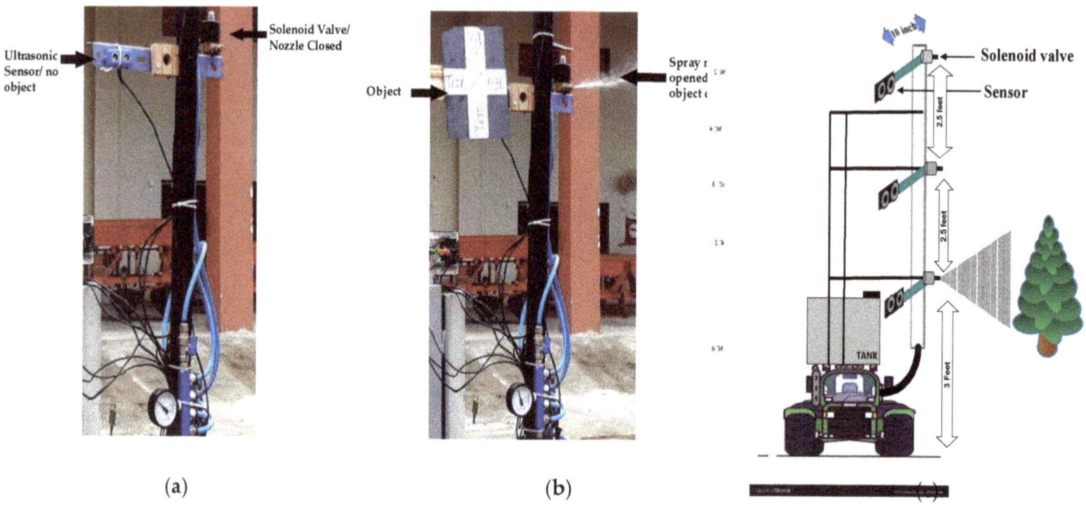

(a) (b) (c)

Figure 1. Laboratory testing of system and design frame. (**a**). No object; hence solenoid valve is closed (**b**). Object detected and solenoid spraying in action (**c**). Design of Spot-Specific sprayer, spraying only when trees are detected, based on their height.

Ultrasonic Sensor

The working of ultrasonic sensors is based on the principle of the Doppler Effect. The Ultrasonic transmitter transmits an ultrasonic sound wave, the signal travels in the air, and when it acquires objected by some material, it gets reflected near the device; this reflected wave is detected through the ultrasonic receiver module [6]. Now, to analyze the distance, we use equation 1; the speed of sound in air is (330 m/s) and the time when the signal is transmitted.

$$\text{Distance} = \text{speed} \times \text{time} \quad (1)$$

This principle is widely used for object detection, obstacle avoidance and depth measurements [7]. Ultrasonic sensors are of two types that are HC-SR04 and JSN-SR04. HR-SR04 has high speed with less distance accuracy, it can detect the object fast, but the measuring distance has (<5%>) error. While the JSN-SR04 has a percentage distance measuring error (<2%>), its detecting speed is slow [8].

The spot-specific sprayer with three solenoid valves and three ultrasonic sensors on a single boom is shown in Figure 1c. This system can be attached to any tractor or small farm vehicle. The ultrasonic sensor will detect the trees, and nozzles attached to the solenoid valves will spray whenever the trees are in the sensing range. Three sensors are attached to 10 feet boom at the height of the citrus tree. If the tree's height is 3 feet, then only the bottom nozzle will start spraying; if the height is 6 feet, then the bottom and middle nozzle will start spraying; and if the tree's height is 9 feet tall, then all three nozzles will start spraying.

3. Results and Discussions

Testing of spot-specific sprayers is divided into two categories. In the first category, the opening and closing intervals of the solenoid valves are calculated for one, two and three sensors separately. The switching interval of the class B (coil insulation) relay is 10 ms, while the power transistor has a 0.1 ms switching interval; both the results of relay and transistor-based switching are shown in Table 1. Due to the fast switching of power transistors, the solenoid valves operate fast, and no part of the tree canopy is missed during spraying.

Table 1. Lab testing results for nozzle opening and closing.

	Transistor-Based Switching		Relay-Based Switching	
Sensor	**Nozzle Opening**	**Nozzle Closing**	**Nozzle Opening**	**Nozzle Closing**
Sensor 1	74 ms	72.2 ms	84 ms	82.2 ms
Sensor 2	71 ms	93.6 ms	81 ms	103.6 ms
Sensor 3	45.8 ms	80.4 ms	55.8 ms	90.4 ms
Sensor 1,2	55 ms	73 ms	65 ms	83 ms
Sensor 2,3	50 ms	58.4 ms	60 ms	68.4 ms
Sensor 1,3	42 ms	79.6 ms	52 ms	89.6 ms
Sensor 1,2,3	28.5 ms	70.4 ms	38.5 ms	80.4 ms

The above results show that the power transistor performs better than relay-based switching as there is a time lag between the opening and closing of solenoid values through the relay, which misses the target tress spraying. The excess amount of agrochemicals is wasted.

Comparison of Uniform Spraying and Spot-Specific Spraying

Spot-specific sprayer provides agrochemical savings of up to 40% for citrus trees. The detail for different agrochemical for citrus and their saving for a 100-acre area is given in Table 2.

Table 2. Amount of different agrochemicals and their savings for spot-specific spraying.

Chemical Name	Used for	Quantity per Acre (mL/L)	Price per Acre	Savings for 100 Acres
Bifenthrin (mL/L)	Leaf Minor	300	1050	42,000/-
Castle (g/L)	Canker	300	1050	42,000/-
Copper Oxychloride (g/L)	Stem disease	250	750	30,000/-

4. Conclusions

The developed prototype of a spot-specific spraying system efficiently detected the objects in laboratory settings through the ultrasonic sensors-based central control system. Before commercialization, field testing is required to fully assess the capability of spotting specific sprayers in tree detection and agrochemical spraying in different orchards. Such a system saves up to 40% of agrochemicals through sensor-based automation, which is economical and easy to implement. All the components are easily available in the market at low prices.

Author Contributions: Conceptualization, S.R.S. and M.N.T.; methodology, S.H. and A.A.; investigation, M.N.T. and S.R.S.; resources, M.N.T.; data curation, S.H. and A.A.; writing—original draft preparation, S.H. and A.A.; writing—review and editing, R.A. and A.A.; visualization, S.H. and R.A.; supervision, S.H. and A.A. All authors have read and agreed to the published version of the manuscript.

Funding: The study is a part of a pilot project for the Data-Driven Smart Decision Platform for Increased Agriculture Productivity (DDSDP) PSDP-funded project No. 332.

Institutional Review Board Statement: Not applicable.

Informed Consent Statement: Not applicable.

Data Availability Statement: The data can be available on request.

Acknowledgments: The authors would like to acknowledge all funding agencies and the scientific and support staff of DDSDP and NCIB projects.

Conflicts of Interest: The authors declare no conflict of interest.

References

1. Waqas, M.S.; Cheema, M.J.M.; Hussain, S.; Ullah, M.K.; Iqbal, M.M. Delayed irrigation: An approach to enhance crop water productivity and to investigate its effects on potato yield and growth parameters. *Agric. Water Manag.* **2021**, *245*, 106576. [CrossRef]
2. Xiao, K.; Ma, Y.; Gao, G. An intelligent precision orchard pesticide spray technique based on the depth-of-field extraction algorithm. *Comput. Electron. Agric.* **2017**, *133*, 30–36. [CrossRef]
3. Gonzalez-de-soto, M.; Emmi, L.; Perez-ruiz, M.; Aguera, J.; Gonzalez-de-santos, P. ScienceDirect Special Issue: Robotic Agriculture Autonomous systems for precise spraying e Evaluation of a robotised patch sprayer. *Biosyst. Eng.* **2016**, *146*, 165–182. [CrossRef]
4. Gangadharan, S.; Burks, T.F.; Schueller, J.K. A comparison of approaches for citrus canopy profile generation using ultrasonic and Leddar® sensors. *Comput. Electron. Agric.* **2019**, *156*, 71–83. [CrossRef]
5. Hussain, S.; Cheema, M.J.M.; Arshad, M.; Ahmad, A.; Latif, M.A.; Ashraf, S.; Ahmad, S. Spray uniformity testing of unmanned aerial spraying system for precise agro-chemical applications. *Pakistan J. Agric. Sci.* **2019**, *56*, 897–903. [CrossRef]
6. Kelemen, M.; Virgala, I.; Kelemenová, T.; Miková, Ľ.; Frankovský, P.; Lipták, T. Distance Measurement via Using of Ultrasonic Sensor. *J. Autom. Control* **2015**, *3*, 71–74. [CrossRef]
7. Derkach, M.; Skarga-bandurova, I. Obstacle Avoidance Algorithm for Small Autonomous Mobile Robot Equipped with Ultrasonic Sensors. In Proceedings of the 2020 IEEE 11th International Conference on Dependable Systems, Services and Technologies (DESSERT), Kyiv, Ukraine, 14–18 May 2020; pp. 236–241.
8. Moussa, M.; Moussa, A. Ultrasonic based heading estimation for aiding land vehicle navigation in gnss denied environment. *Int. Arch. Photogramm. Remote Sens. Spat. Inf. Sci.* **2018**, *XLII*, 10–12. [CrossRef]

environmental sciences proceedings

MDPI

Proceeding Paper

Feasibility of Ultrasonic Sensors in Development of Real-Time Plant Canopy Measurement System †

Abdul Kaleem [1], Muhammad Aqib [1,2,*], Shoaib Rashid Saleem [3] and Muhammad Jahanzeb Masud Cheema [1,4]

1 National Center of Industrial Biotechnology, PMAS Arid Agriculture University Rawalpindi, Rawalpindi 46000, Pakistan
2 University Institute of Information Technology, PMAS Arid Agriculture University Rawalpindi, Rawalpindi 46000, Pakistan
3 Department of Farm Machinery and Precision Engineering, PMAS Arid Agriculture University Rawalpindi, Rawalpindi 46000, Pakistan
4 Department of Land and Water Conservation Engineering, PMAS Arid Agriculture University Rawalpindi, Rawalpindi 46000, Pakistan
* Correspondence: aqib.qazi@uaar.edu.pk
† Presented at the 1st International Precision Agriculture Pakistan Conference 2022 (PAPC 2022)—Change the Culture of Agriculture, Rawalpindi, Pakistan, 22–24 September 2022.

Abstract: A real-time approach based on IoT sensors to detect and measure the canopy size of trees in orchards for plant data collection has been proposed. This work discusses the issues related to sensors, particularly ultrasonic sensors for canopy size measurement. Other core issues related to sound cone measurement, angle error, crosstalk error, and measurement accuracy have also been investigated in depth. Keeping these aspects in mind, this work focused on the usability of these sensors while providing information about environmental structures. The feasibility of this research was tested in a laboratory. The results showed that for large sensor spacings, the interference errors are minimal, and the sensors' field distance measurements are accurate.

Keywords: IoT; digital agriculture; precision agriculture; ultrasonic sensor; plant canopy measurement; error reductions

Citation: Kaleem, A.; Aqib, M.; Saleem, S.R.; Cheema, M.J.M. Feasibility of Ultrasonic Sensors in Development of Real-Time Plant Canopy Measurement System. *Environ. Sci. Proc.* **2022**, *23*, 22. https://doi.org/10.3390/environsciproc2022023022

Academic Editor: Saddam Hussain

Published: 26 December 2022

1. Introduction

Numeral and visual research of top of the plant arrangements has massive efficacy for phenotypical findings. In such studies, users can automatically obtain information with pragmatic measurements. The improved capacity of techniques in computer processing and the reduced size of recent data measurement devices has supported an exponential increase in plant canopy measurements [1]. Different IoT sensors are used for estimating plant qualities [2]. Traditional application of pesticides has resulted in a drift due to the employment of continuous and non-selective spraying methods without adequate control mechanisms. To counter this drift, Precise Variable Spraying Technology is employed to lessen the effect of pesticide waste and environmental pollution. As mentioned, technology supports automatically adjustable nozzle flow rate, the volume of air supplied, and variable nozzle—tree distances depending upon the canopy characteristics [3–5]. This effective use of precision technology results in an increased productivity and reduced costs of inputs. Electronic sensors have different advantages and drawbacks that depend upon their acquiring cost, processing speed, and data size.

In [6], plant reconstruction was estimated with sonar intensity to measured plant volume density. A cylindrical leaf-distribution canopy model based upon the experimental results was proposed in [7]. Apart from ultrasonic sensor-based detection and estimation systems, researchers have shown interest in and have applied methods for tree canopy detection, using laser sensors and LIDAR. In comparison, LIDAR provides better estimates of the crop variables than ultrasonic sensors. The researchers in [8,9] suggested a real-time canopy quantity recognition system by applying a field laser sensor and inferred the undergrowth external base from top capacity evidence and statistics. Evidently, detection technologies are different in terms of their various inherent characteristics. For example, LIDAR-based detection systems demand high initial costs, are intrinsically complex, and result in large datasets that require large computational resources [10]. Conversely, ultrasonic sensor-based detection mechanisms have a low initial cost, are simple in terms of application, and are more practical in various environments. This paper discusses the general process of canopy reconstruction of plants. Then, the technical defects and reasons for low accuracy are outlined. The sensor's applicability in terms of sound-cone measurement, angle error reduction, crosstalk error reduction, and field measurement accuracy are also assessed in this study.

2. System Model

Ultrasonic sensors involve two modules, one acting as the transmitter and the other acting as the receiver. The transmitter will transform the electrical signal into a high-frequency ultrasonic pulse and the receiver is responsible for receiving the signal that bounces back. When the receiver detects the sonar pulse, it will generate an output signal that is proportional to the magnitude of the distance of the object. The pulses are triggered using microcontrollers; as a result, the ultrasonic transmitter emits a burst of pulses having a frequency of about 40,000 Hz as shown in Figure 1. After transmitting the signal, the receiver is activated for a certain time (38 milliseconds in this study). The calculation of the distance between the object and the sensor is carried out with a microcontroller using following Equation [11].

$$s = txv \tag{1}$$

where 's' is the distance of the object, 't' is the time or width of the pulse sensed by the receiver, and 'v' is the speed of sound. Equation (1) calculates the complete distance traveled by the pulse after transmission and reception at the receiver, which is double the distance between the sensor and object. To find the exact distance of the object, Equation (1) can be written as:

$$s = (txv)/2 \tag{2}$$

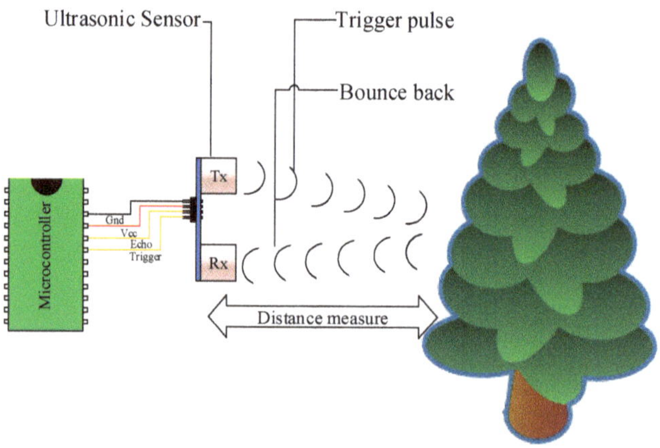

Figure 1. Working principle of the ultrasonic sensor.

Using this feature the plant/tree canopy is recognized using an array of ultrasonic sensors. Multiple sensors are used to gather samples of plant depth at different heights. The microcontroller is continuously reading the data from the sensors so the data from each ultrasonic sensor 'A_A' at the given instance is averaged to determine the cross-sectional canopy of the plants [12].

$$P_c = 2\sum_{j=1}^{j=10}\left(0.5A_R - \overline{A}_{Aj}\right)A_s \quad (3)$$

In Equation (3), 'P_c' is the plant canopy in square meters, A_R is the reference distance, $\overline{A}_{Aj}$ is the jth average distance for a specific time interval, and A_s is the displacement between the ultrasonic sensors. Equation (3) calculates the single cross-section of a plant at a certain instance. The overall canopy of plants is calculated using:

$$v_{PC} = \sum_{j=1}^{j=i} tW_jP_{cj} \quad (4)$$

where v_{PC} is the plant canopy in cubic meters, i is the number of scans or instances in which multiple plant scans are carried out, 't' is the time interval for one scan, W_j is the jth scan for canopy calculation, and P_{cj} is the jth cross section of the plant. The canopy of the plants in our prototype is calculated using Equation (4). Whereas some variables must be set directly, GPS and wheel transponders can be changed in later versions. This paper's research objective was to assess the viability of using a sensor module to calculate the canopy.

3. Results and Discussion

The accuracy of ultrasonic sensors is recognized in this study. Multiple tests related to ultrasonic sensors was conducted to examine their accuracy, which is dependent on distance and the angle of obstacles and sensors. Increasing the number of sensors will decrease the spacing between the adjacent sensors and as a result, interfacing will occur. Table 1 shows the comparison between different sensor spacing distances for the ultrasonic sensors.

Table 1. Comparison between multiple ultrasonic sensors spacing distances.

Object Distance (cm)	Sensor Spacing (cm)			
	30	60	75	90
25	NI	NI	NI	NI
43	I	NI	NI	NI
73	I	I	NI	NI
84	I	I	I	NI
98	I	I	I	I

NI = No Interference and I = Interference.

This method of signal sensing will not obtain accurate information at a given point. The thickness of the ultrasonic signal increases as the signal travels a long distance. As the beam widens, it will bounce back off the first surface that it contacts as shown in Figure 2. This error is dependent upon the shape of obstacle and the distance from sensor.

The sensor angle is also an important parameter for consideration in canopy measurements. This parameter was also tested by changing the spacing of the object and sensor placement angle as shown in Table 2. It was observed that the chosen sensor has satisfactory performance up to 15 degrees at any distance.

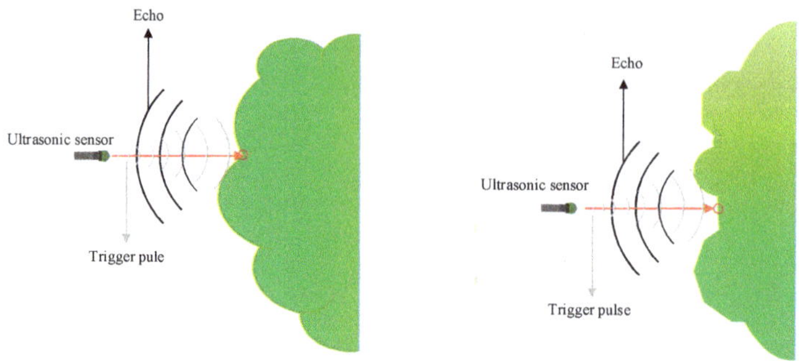

Figure 2. Beam widening of the ultrasonic sensor.

Table 2. Comparison of ultrasonic sensors angle of detection.

Angle	Actual Distance (cm)	Measure Distance (cm)	Error (cm)
0°	55	52	2
	105	108	3
	150	156	6
10°	55	52	2
	105	108	3
	150	155	5
15°	55	53	3
	105	109	4
	150	158	8
20°	55	Out of range	Out of range
	105	Out of range	Out of range
	150	Out of range	Out of range

4. Conclusions

Agriculture has advanced using IoTs for crop management and optimization. In such studies, users can automatically obtain information with pragmatic measurements. The improved capacity of techniques in computer processing has supported an exponential increase in plant canopy measurements and reconstruction studies. Ultrasonic sensor-based detection mechanisms have a low initial cost, are simple in terms of application, and are more practical in various environments. These sensors' applicability in terms of sound-cone measurement, angle error, crosstalk error, and field measurement accuracy are important parameters to observed.

Author Contributions: Conceptualization, M.A. and S.R.S.; methodology, A.K.; investigation, A.K.; resources, M.A., S.R.S. and M.J.M.C.; data curation, A.K. and M.A.; writing—original draft preparation, A.K.; writing—review and editing, M.A. and S.R.S.; visualization, A.K.; supervision, M.A., S.R.S. and M.J.M.C. All authors have read and agreed to the published version of the manuscript.

Funding: This study is a part of a pilot project for the Establishment of a National Center of Industrial Biotechnology for Pilot Manufacturing of Bioproducts Using Synthetic Biology and Metabolic Engineering Technologies (ENCIB) PSDP-funded project No. 321.

Institutional Review Board Statement: Not applicable.

Informed Consent Statement: Not applicable.

Data Availability Statement: Not applicable.

Acknowledgments: The authors would like to acknowledge all funding agencies and the scientific and support staff of the NCIB project.

Conflicts of Interest: The authors declare no conflict of interest.

References

1. Vázquez-Arellano, M.; Griepentrog, H.W.; Reiser, D.; Paraforos, D.S. 3-D imaging systems for agricultural applications—A review. *Sensors* **2016**, *16*, 618. [CrossRef] [PubMed]
2. Blais, F. Review of 20 years of range sensor development. *J. Electron. Imaging* **2004**, *13*, 231–240. [CrossRef]
3. Nan, Y.; Zhang, H.; Xu, Y.; Jiao, X.; Zheng, J.; Liu, D. Research progress on profiling target spray and its control technology in agriculture and forestry. *World For. Res.* **2018**, *31*, 54–58.
4. Jeon, H.Y.; Zhu, H. Development of a variable-rate sprayer for nursery liner applications. *Trans. ASABE* **2012**, *55*, 303–312. [CrossRef]
5. Gil, E.; Escolà, A.; Rosell, J.R.; Planas, S.; Val, L. Variable rate application of plant protection products in vineyard using ultrasonic sensors. *Crop Prot.* **2007**, *26*, 1287–1297. [CrossRef]
6. Palleja, T.; Landers, A.J. Real time canopy density validation using ultrasonic envelope signals and point quadrat analysis. *Comput. Electron. Agric.* **2017**, *134*, 43–50. [CrossRef]
7. Nan, Y.; Zhang, H.; Zheng, J.; Bian, L. Estimating leaf area density of Osmanthus trees using ultrasonic sensing ScienceDirect Estimating leaf area density of Osmanthus trees using ultrasonic sensing. *Biosyst. Eng.* **2019**, *186*, 60–70. [CrossRef]
8. Palacín, J.; Pallejà, T.; Tresanchez, M.; Sanz, R.; Llorens, J.; Ribes-Dasi, M.; Masip, J.; Arno, J.; Escola, A.; Rosell, J.R. Real-time tree-foliage surface estimation using a ground laser scanner. *IEEE Trans. Instrum. Meas.* **2007**, *56*, 1377–1383. [CrossRef]
9. Palacin, J.; Palleja, T.; Tresanchez, M.; Teixido, M.; Sanz, R.; Llorens, J.; Arno, J.; Rosell, J.R. Difficulties on tree volume measurement from a ground laser scanner. In Proceedings of the 2008 IEEE Instrumentation and Measurement Technology Conference, Victoria, BC, Canada, 12–15 May 2008; pp. 1997–2002.
10. Deepak, B.B.V.L.; Bahubalendruni, M.V.A.R.; Biswal, B.B. Development of in-pipe robots for inspection and cleaning tasks: Survey, classification and comparison. *Int. J. Intell. Unmanned Syst.* **2016**, *4*, 182–210. [CrossRef]
11. Morgan, E.J. HC—SR04 Ultrasonic Sensor Datasheet. Available online: https://docplayer.net/23967545-Hc-sr04-ultrasonic-sensor-elijah-j-morgan-nov.html (accessed on 20 December 2022).
12. Schumann, A.W.; Zaman, Q.U. Software development for real-time ultrasonic mapping of tree canopy size. *Comput. Electron. Agric.* **2005**, *47*, 25–40. [CrossRef]

environmental sciences proceedings

MDPI

Proceeding Paper

Evaluating the Impact of the Billion Tree Afforestation Project (BTAP) on Surface Water Flow in Tarbela Reservoir Using SWAT Model †

Abdul Basit [1,2,*], Abid Sarwar [2], Saddam Hussain [2,3], Shoaib Saleem [1,4], Basit Raza [1], Muhammad Ali Hassan Khan [5] and Muhammad Abubakar Aslam [2,6]

1 Green AI, Center of Precision Agriculture, PMAS Arid Agriculture University, Rawalpindi 46000, Pakistan
2 Department of Irrigation and Drainage, University of Agriculture Faisalabad, Faisalabad 38000, Pakistan
3 Department of Agricultural and Biological Engineering, University of California (UC Davis), Davis, CA 95616, USA
4 Department of Farm Machinery and Precision Engineering, Faculty of Agricultural Engineering and Technology, PMAS Arid Agricultural University, Rawalpindi 46000, Pakistan
5 Department of Structural Engineering, University of Agriculture Faisalabad, Faisalabad 38000, Pakistan
6 NCIB, Center of Precision Agriculture, PMAS Arid Agriculture University, Rawalpindi 46000, Pakistan
* Correspondence: agrarian230@gmail.com; Tel.: +92-3029629385
† Presented at the 1st International Precision Agriculture Pakistan Conference 2022 (PAPC 2022)—Change the Culture of Agriculture, Rawalpindi, Pakistan, 22–24 September 2022.

Abstract: Khyber Pakhtunkhwa launched the Billion Tree Afforestation Project (BTAP) in 2014. Pakistan also initiated a "10 billion trees in five years" project in 2018. The soil and water assessment tool (SWAT) model was used to forecast the impacts of LULC changes on water yield under three scenarios: before planting, after 1 billion trees planted, and after 10 billion trees planted. Model calibration and validation were undertaken at the Bisham Qila gauging station from 1984 to 2000 and 2001 to 2010. The Tarbela reservoir's mean annual runoff declined from 53.70 mm to 45.40 mm after 1 billion trees planted, while under the third scenario it approximated 35.05 mm.

Keywords: billion tree project; Terbela reservoir; SWAT model; sediment load; water yield

Citation: Basit, A.; Sarwar, A.; Hussain, S.; Saleem, S.; Raza, B.; Khan, M.A.H.; Aslam, M.A. Evaluating the Impact of the Billion Tree Afforestation Project (BTAP) on Surface Water Flow in Tarbela Reservoir Using SWAT Model. *Environ. Sci. Proc.* **2022**, *23*, 23. https://doi.org/10.3390/environsciproc2022023023

Academic Editor: Muhammad Jehanzeb Masud Cheema

Published: 26 December 2022

1. Introduction

Surface runoff with sediment is caused by erosion. Precipitation (snowfall and rain) and land cover affect soil erosion and runoff [1]. Urbanization increases runoff and sedimentation, reducing infiltration [2]. In the Tarbela reservoir, surface runoff has the greatest impact on sedimentation [3]. Land use change affects runoff and sediment flow. Khyber Pakhtunkhwa (KPK) launched the Billion Tree Afforestation Project (BTAP) in 2014 to meet the country's water needs and to mitigate and adapt to climate change. On 3 September 2018, Pakistan's prime minister announced the planting of 10 billion plants across the country, 3 billion of which will be planted in the upper Indus Basin. By 2024, 10 BTAP projects will be completed [4]. We hypothesized that BTAP plantations would reduce the Indus River surface water flow.

There have been no published studies that described the surface runoff at Tarbela reservoir after BTAP planting. We used the SWAT model to calculate the effect of LULC on streamflow. We used three LULC datasets to simulate the SWAT model. Before BTAP planting, stream flow was determined using LULC data. The second scenario used 1 billion trees' LULC data. In the third scenario, LULC was used after 10 billion trees were planted.

2. Materials and Methods

2.1. Study Area

The study area was the Tarbela Dam drainage basin, one of Pakistan's largest reservoirs [5]. Terbela Dam is situated at 34°05′23″ N and 72°41′54″ E on the Indus River in Haripur and Swabi. It is 143 m high [4].

2.2. SWAT Model

To simulate hydrological processes, SWAT needs climate variables, LULC, soil data, and river basin management [6]. SWAT can be discretized by grid or aggregated, in addition to HRU [7]. This study used HRU-based discretization. Figure 1 shows the method's flowchart.

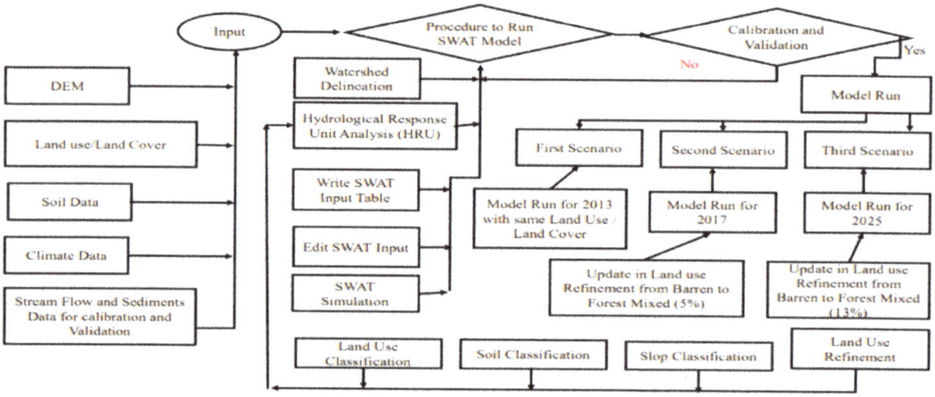

Figure 1. Workflow chart of methodology [8].

2.3. LULC Scenarios

In land use refinement, land use classes were updated according to changes in the study area, from barren to forest cover. After recreating HRU, the entire forest in the study area expanded as land use was updated. In scenarios 2 and 3, this tool changed barren land into forest. The second scenario was considered from after 1 billion trees were planted until 2017. In scenario 3, 10 billion trees will be planted by 2024, of which 3 billion will be in our study area [4]. In the second scenario, after planting 1 billion trees, 5% BARR land became 5% forest mixed, and in the third scenario, 13% barren land became 13% forest-mixed (FRST); these land use percentages were updated, respectively, in land use refinement.

3. Results

3.1. Calibration and Validation

In this study, calibration and validation of the SWAT model were done at Bisham Qila station. Monthly calibration occurred from 1984 to 2000, and monthly validation from 2001 to 2010. SWAT-CUP SUFI2 was used for calibration. The calibration results were promising, indicating a high model performance that can be used to examine land use and land cover effects on sediments and stream flow.

Figure 2 and Table 1 show the calibration and validation results. The calibration and validation values for R2, PBIAS, and NSE ranged from 0.84 to 0.88.

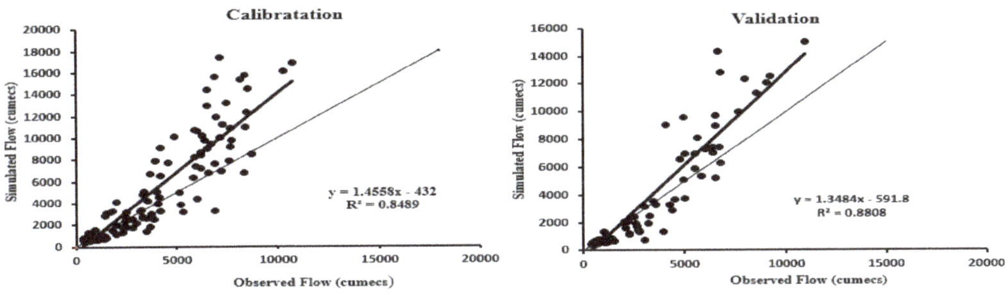

Figure 2. SWAT model calibration and validation results.

Table 1. The calibration and validation values.

Performance Indicator	Calibration	Validation (Water Yield)
R^2	0.85	0.88
PBIAS	11.2	9.4
NSE	0.84	0.86

3.2. Impacts of BTAP on Water Yield

Figure 3 shows SWAT sub-basin water yields for 2013, 2017, and 2025. It includes model-developed water yield components. The first, second, and third sub-basin scenarios yielded 54 mm, 45 mm, and 35 mm of water annually, respectively.

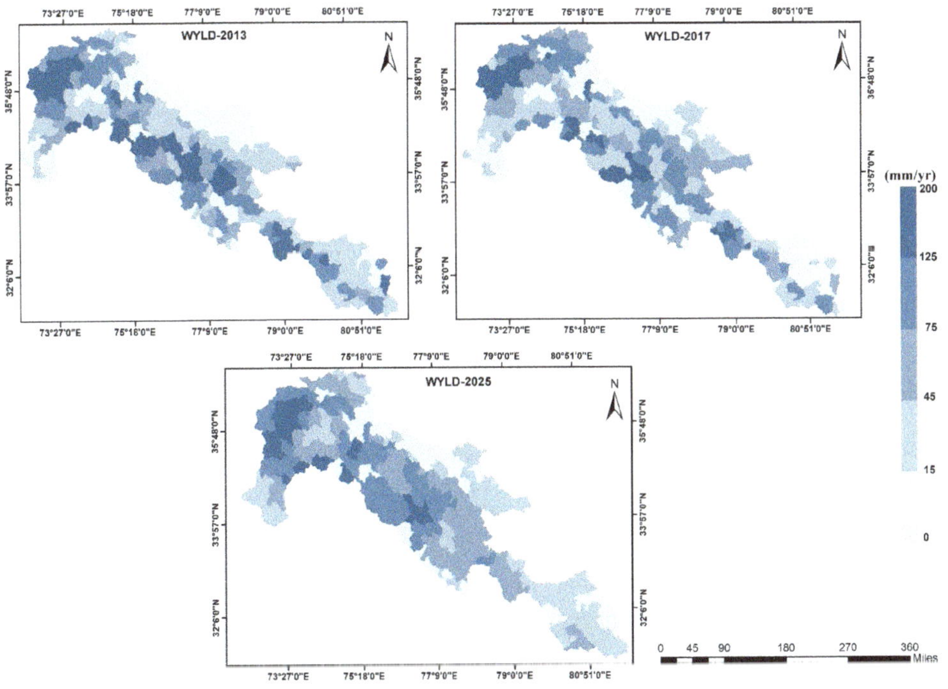

Figure 3. Spatial distribution of water yield for three scenarios (2013, 2017, and 2025) in sub-basins.

Figure 4 shows the model results compared for accumulated water yield at the last sub-basin (300), an inlet of the Tarbela reservoir. After planting 1 billion trees, the peak monthly flow decreased from 14,300 m^3/s (in 2013) to 14,225 m^3/s (in 2017). After 10 billion trees planted, peak flow was reduced by 12,100 m^3/s. Increased forestation from 2013 to 2017 reduced the annual flow by 18%. SWAT evaluation indicates that a 13% rise in forest cover would cause annual flows to drop by 26% from 2013 to 2025.

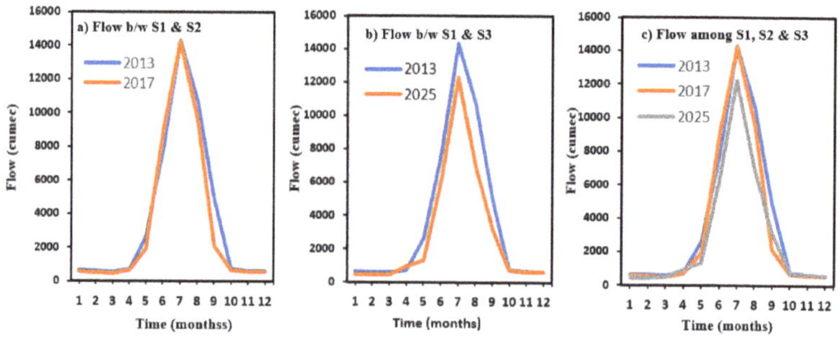

Figure 4. Monthly stream flow comparison between first, second, and third scenarios.

4. Conclusions

Simulations showed that increasing tree planting reduces water flow. A 3% increase in forest cover (after one billion trees) reduced reduce surface water flow by 18%, and a 13% increase (after three billion trees) reduced water flow by 26%, and sediment flow also reduced, respectively. The effect of BTAP on runoff shows that the Tarbela Dam area needs more forest cover. To stop runoff and sediment from getting into reservoirs, it is suggested that the Indus watershed be managed properly.

Author Contributions: Conceptualization, methodology, software, validation, A.B., A.S., and S.H.; formal analysis, S.S. and M.A.A.; investigation, A.S.; resources, A.B. and S.H.; data curation, A.B. and B.R.; writing—original draft preparation, A.B.; writing—review and editing, S.H. and M.A.H.K.; visualization, S.S.; supervision, A.S. and M.A.A.; project administration, S.H. and S.S. All authors have read and agreed to the published version of the manuscript.

Funding: This research received no external funding.

Institutional Review Board Statement: Not Applicable.

Informed Consent Statement: Not Applicable.

Data Availability Statement: Not Applicable.

Conflicts of Interest: The authors declare no conflict of interest.

References

1. Salam, M.; Cheema, M.J.M.; Zhang, W.; Hussain, S.; Khan, A.; Bilal, M.; Zaman, M.A. Groundwater storage change estimation using grace satellite data in Indus Basin. *Big Data Water Resour. Eng. (BDWRE)* **2020**, *1*, 13–18. [CrossRef]
2. Iqbal, M.M.; Li, L.; Hussain, S.; Lee, J.L.; Mumtaz, F.; Elbeltagi, A.; Dilawar, A. Analysis of Seasonal Variations in Surface Water Quality over Wet and Dry Regions. *Water* **2022**, *14*, 1058. [CrossRef]
3. Dahri, Z.H.; Ahmad, B.; Leach, J.H.; Ahmad, S. Satellite-based snowcover distribution and associated snowmelt runoff modeling in Swat River Basin of Pakistan. *Proc. Pak. Acad. Sci.* **2011**, *48*, 19–32.
4. Mirza, U.K.; Ahmad, N.; Majeed, T.; Harijan, K. Hydropower use in Pakistan: Past, present and future. *Renew. Sustain. Energy Rev.* **2008**, *12*, 1641–1651. [CrossRef]
5. Getachew, H.E.; Melesse, A.M. The impact of land use change on the hydrology of the Angereb Watershed, Ethiopia. *Int. J. Water Sci.* **2012**, *1*. [CrossRef]
6. Savvidou, E. A study of alternative Hydrological Response Units (HRU) Configurations in the Context of Geographical Information Systems (GIS)-Based Distributed Hydrological Modeling. Doctoral Dissertation, Department of Civil Engineering and Geoinformatics Engineering, School of Engineering and Technology, Cyprus University of Technology, Limassol, Cyprus, 2018.

7. Shafeeque, M.; Sarwar, A.; Basit, A.; Mohamed, A.Z.; Rasheed, M.W.; Khan, M.U.; Sabir, R.M. Quantifying the Impact of the Billion Tree Afforestation Project (BTAP) on the Water Yield and Sediment Load in the Tarbela Reservoir of Pakistan Using the SWAT Model. *Land* **2022**, *11*, 1650. [CrossRef]
8. Dilawar, A.; Chen, B.; Ashraf, A.; Alphonse, K.; Hussain, Y.; Ali, S.; Hussain, S. Development of a GIS based hazard, exposure, and vulnerability analyzing method for monitoring drought risk at Karachi, Pakistan. *Geomat. Nat. Hazards Risk* **2022**, *13*, 1700–1720. [CrossRef]

environmental sciences proceedings

MDPI

Proceeding Paper

Regionalization of Drought across Pakistan †

Tausif Khan [1], Zeeshan Waheed [1], Muhammad Nauman Altaf [1], Muhammad Naveed Anjum [1], Fiaz Hussain [1] and Muhammad Azam [2,*]

[1] Faculty of Agricultural Engineering and Technology, PMAS Arid Agriculture University, Rawalpindi 46000, Pakistan
[2] Department of Structures and Environmental Engineering, PMAS Arid Agriculture University, Rawalpindi 46000, Pakistan
* Correspondence: muhammad.azam@uaar.edu.pk

† Presented at the 1st International Precision Agriculture Pakistan Conference 2022 (PAPC 2022)—Change the Culture of Agriculture, Rawalpindi, Pakistan, 22–24 September 2022.

Abstract: Due to Pakistan's complex hydro-climatic and topographical features, drought is a severe problem. It is necessary to regionalize various topographical and hydrometeorological occurrences into uniform zones. The regionalization of clusters across Pakistan has been examined and analyzed using the hierarchical classification of principal components (HCPC). Five statistically homogenous zones were made, which were validated through the cluster validation indices. Univariate discordancy tests were run using the drought's severity and duration as inputs. Over 12 months, drought was regionalized for SPEI time scales, indicating regional discordancy in cluster 4, while cluster 2 had a smaller number of stations, which were further adjusted to ensure homogeneity. The results of this research might be utilized to offer the fundamental information needed to develop a regional drought mitigation plan.

Keywords: HCPC; regionalization; SPEI; univariate; homogeneity; discordancy test

Citation: Khan, T.; Waheed, Z.; Altaf, M.N.; Anjum, M.N.; Hussain, F.; Azam, M. Regionalization of Drought across Pakistan. *Environ. Sci. Proc.* **2022**, *23*, 24. https://doi.org/10.3390/environsciproc2022023024

Academic Editor: Muhammad Naveed Tahir

Published: 26 December 2022

1. Introduction

Since its inception, Pakistan has endured droughts, on average, four out of every ten years, making them a frequent thing among significant catastrophic events. In 1998, Pakistan seems to have had its worst dry spell since 1947, which persisted until 2002 [1]. Weather patterns are directly linked to meteorological drought, which often results from inadequate precipitation over a region [2].

The country has experienced drought as a result of the highest evaporation and transpiration due to global warming. As a result, it is possible to classify the coupled behavior of elements, such as environmental, physical geography, and hydrological features, and to relate drought to factors that are not necessarily independent. In regional modeling analysis, multivariate strategies refer to a group of methods, including hydro-meteorological, climatic, and physiographic factors that must be intimately intertwined.

With a standardized precipitation and evapotranspiration index as the foundation for the k cluster analysis and tree edge removal strategies, drought regionalization was done in Pakistan from 1902 to 2015 [1].

Using PCA, the primary factors causing the deterioration in drinking water quality in Gilgit, northern Pakistan, were investigated [3]. The proper identification of contiguous zones permits a parametric estimate of drought frequency in a regional study, which may be utilized for planning and managing regional drought risk.

The rest of the studies have concentrated on topographical, climatic, and drought factors using at-site (local) multivariate analysis. On the other hand, there have not been a lot of studies performed on the local representation of drought modeling at ungauged locations. The main goal of the project is to regionalize drought using L-moments.

2. Materials and Methods

We have taken the total number of 41 rain gauge stations in Pakistan covering the whole country. The flow chart of our methodology is shown in Figure 1.

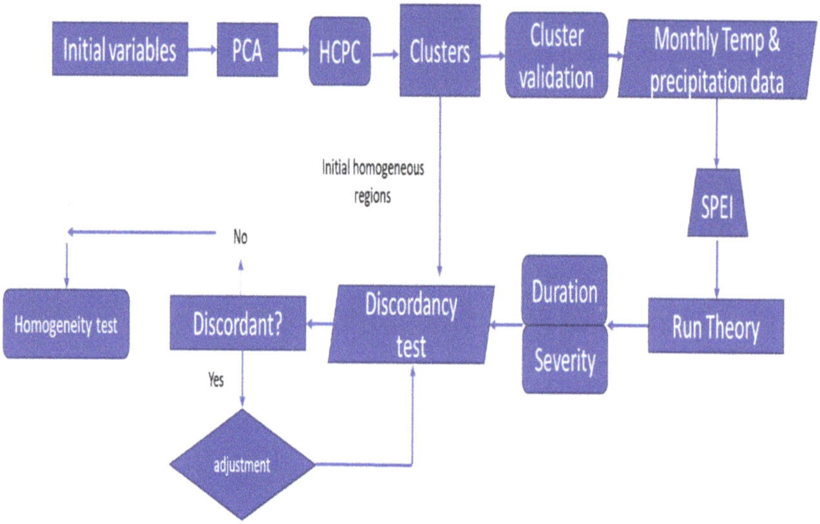

Figure 1. Study flowchart.

3. Results

3.1. Principal Component Analysis

Each of the eight variables were normalized to have a standard deviation of one and a mean of zero. The inter-correlation matrix revealed a 0.54 correlation coefficient between mean annual precipitation (MAP) and latitude (LAT), and 0.50 between MAP and longitude (Long). Whereas latitude (LAT) and elevation are shown to have a 0.55 correlation coefficient, mean daily maximum temperature (MDMXT) and mean daily minimum temperature (MDMNT) show the strongest correlation with evapotranspiration (ET^0) of about 0.8 and 0.79. Elevation (Ele), along with mean annual precipitation (MAP), longitude (Long), and relative humidity (RH), shows a moderate correlation of about 0.16 to 0.3.

3.2. Variances of Each Principal Component (PC)

The correlation matrix is used for principal component analysis. The first four principal components (PC) retained 95.2 percent of the information (variances). The first, second, third and fourth principal component retained 57.9%, 18.6%, 10.6%, and 8.1% of the information (variances) respectively. So, therefore, the major components have been chosen for cluster analysis.

The average contribution of the variables is 12.5%. A contribution higher than this is considered important in contributing to the components. Long and mean annual precipitation (MAP) contributed significantly to the second major component. RH contributed significantly to the third principal component. MAP and LONG contributed significantly to the fourth principal component. Those less important variables for understanding component variability are often excluded from evaluation in the PCA.

3.3. Hierarchal Clustering on Principal Components

The PCA data were analyzed using the HCPC approach. Clusters were initially formed by the technique of HCPC while mapping Pakistan's geographical space to each location.

Cluster validation indices were used to aid in determining the optimal number of clusters.

3.4. Validation of Cluster Indices

Four cluster validity indices were tested to find the best group of clusters and varying permissible cluster sizes. The silhouette (S) index, the Dunn index, and the Calinski and Harabasz index all maximize their values. When there are five clusters, the DB index becomes minimal [4]. Therefore, when there were five clusters, the values of all validity indices indicated good correlations.

3.5. SPEI-12 Computation

SPEI-12 was estimated for all metrological stations in Pakistan from 1981 to 2018 in a 12-year time frame in this study. The run theory was used to quantify that "the highest mean severities and durations have been found at Badin and Jiwani stations, which are 54.93 and 33 months respectively" [5].

3.6. Discordancy and Heterogeneity Measures

By using "L-moments techniques", discordancy and heterogeneity values were measured using R-programming. It is decided that cluster 4 has only one station, Bahawalnagar, which is "discordant" and is shifted to cluster 5 of the study area. On the other hand, there was a smaller number of stations than 5. As it is impossible for the clustering algorithm to decide the connectivity and similarities within the cluster when the number of observations is less than 5. So, after shifting the Panjgur station from cluster 5 to cluster 3, the codes were run again, and, finally, we have got the homogeneous clusters, which can be seen in the following cluster map of Pakistan.

L-moments approaches have been performed to examine discordancy and heterogeneity based on the nation's physiographic and climatic parameters [6]. The cluster analysis findings reveal wide variation in the sites of the three groups, which are largely mountainous, but a strong similarity in the other two. The summary of "L-moments statistics" is given in Table 1 while the clusters which have been evaluated show only discordancy at cluster 4.

Table 1. Characteristics of clusters evaluated by L-moments discordancy and heterogeneity tests.

S. No	Zones	No of Stations	Discordant Sites	H_{1D}	Homogenous/Heterogeneous
1	Cluster-1	Chilas, Gilgit, Chitral, Saidu-Sharif, Gupis, Murre, Parachinar, Dir	0	−0.7	Homogenous
2	Cluster-2	Jiwani, Karachi, Badin, Pasni	0	1.32	Homogenous
3	Cluster-3	Dalbandin, Nokkundi, Khuzdar, Barkhan, Zhob, Kalat	0	0.14	Homogenous
4	Cluster-4	Bahawalnagar, Bahawalpur, Khanpur, Mulatan, Sibbi, Nawab Shah, Padidian, Jacobabad, Rohri	1	1.19	Heterogeneous
5	Cluster-5	Kotli, Lahore, Jhelum, Faisalabad, Sargodha, Islamabad, Sialkot, Miawali, D.I khan, Kohat, Peshawar	0	0.25	Homogenous

We have performed shifting and adjusted these statistical measurements that verified the acceptable homogeneity of the five altered zones. As a result, five homogenous clusters for the study area Pakistan are shown in the Figure 2.

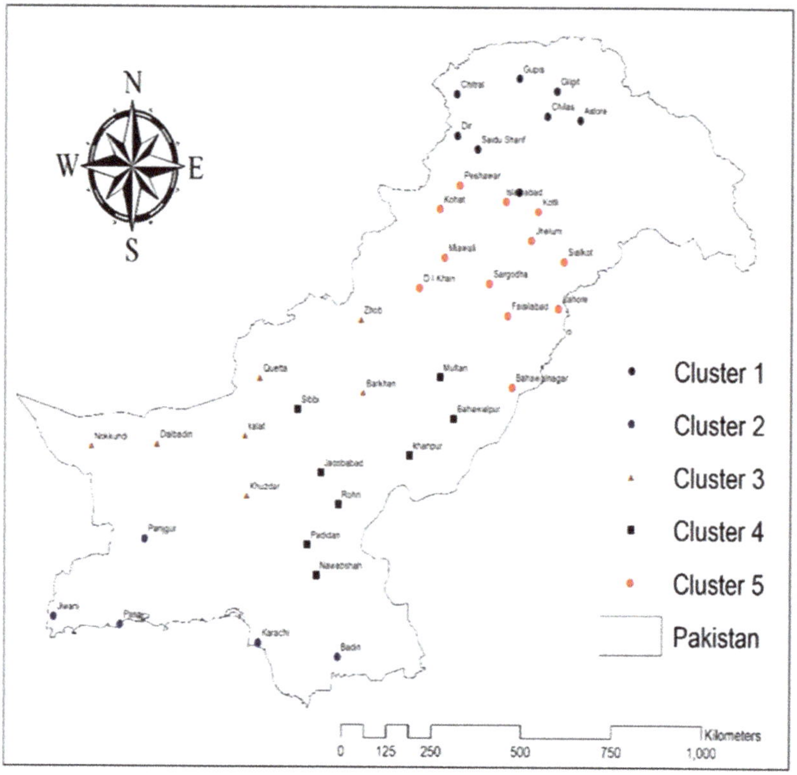

Figure 2. Homogenous clusters map for the study area of Pakistan.

4. Conclusions

The purpose of this study was to look at the regionalization of drought characteristics in conjunction with other physiographic and climate variables. It seeks to deal with Pakistan's complicated hydro climatic and topographical aspects. As a result, eight essential hydrologic, climatic, and physiographic criteria were chosen for the regionalization process. Drought characteristics were collected from 41 rainfall sites using the SPEI truncation level technique. The HCPC algorithm, a hybrid of Ward's classification method, the K-means algorithm, and the PCA methodology, is explored and applied for drought regionalization in Pakistan. The cluster produced by the HCPC technique was used to calculate discordancy and homogeneity for the SPEI-12-time scale.

Author Contributions: All authors contributed equally. All authors have read and agreed to the published version of the manuscript.

Funding: This research received no external funding.

Institutional Review Board Statement: Not applicable.

Informed Consent Statement: Written informed consent has been obtained from the patient(s) to publish this paper.

Data Availability Statement: Not applicable.

Conflicts of Interest: The authors declare no conflict of interest.

References

1. Jamro, S.; Dars, G.H.; Ansari, K.; Krakauer, N.Y. Spatio-temporal variability of drought in Pakistan using standardized precipitation evapotranspiration index. *Appl. Sci.* **2019**, *9*, 4588. [CrossRef]
2. Kao, S.C.; Govindaraju, R.S. A copula-based joint deficit index for droughts. *J. Hydrol.* **2010**, *380*, 121–134. [CrossRef]
3. Fatima, S.U.; Khan, M.A.; Siddiqui, F.; Mahmood, N.; Salman, N.; Alamgir, A.; Shaukat, S.S. Geospatial assessment of water quality using principal components analysis (PCA) and water quality index (WQI) in Basho Valley, Gilgit Baltistan (Northern Areas of Pakistan). *Environ. Monit. Assess.* **2022**, *194*, 151. [CrossRef] [PubMed]
4. Hosking, J.R. L-moments: Analysis and estimation of distributions using linear combinations of order statistics. *J. R. Stat. Soc. Ser. B* **1990**, *52*, 105–124. [CrossRef]
5. Rini, D.S.; Novianti, P.; Fransiska, H. Internal cluster validation on earthquake data in the province of Bengkulu. *IOP Conf. Ser. Mater. Sci. Eng.* **2018**, *335*, 012048. [CrossRef]
6. Ullah, H.; Akbar, M.; Khan, F. Construction of homogeneous climatic regions by combining cluster analysis and L-moment approach based on the Reconnaissance Drought Index for Pakistan. *Int. J. Climatol.* **2020**, *40*, 324–341. [CrossRef]

environmental sciences proceedings

MDPI

Proceeding Paper

Evaluation of the Water Distribution Efficiency of Wheat under Improved Water and Fertilizer Application Techniques †

Lubna Anjum [1,*], Qamar Rasheed [1], Muhammad Rizwan [2], Abdul Rehman [1] and Muhammad Shahid Ibni Zamir [3]

1 Department of Irrigation and Drainage, University of Agriculture Faisalabad, Faisalabad 38000, Pakistan
2 Department of Civil Engineering, Swedish College of Engineering and Technology, Rahim Yar Khan 64200, Pakistan
3 Department of Agronomy, University of Agriculture Faisalabad, Faisalabad 38000, Pakistan
* Correspondence: lubnaanjum01@gmail.com
† Presented at the 1st International Precision Agriculture Pakistan Conference 2022 (PAPC 2022)—Change the Culture of Agriculture, Rawalpindi, Pakistan, 22–24 September 2022.

Abstract: The application of water-retention polymers with improved fertilizers is a better crop-growing technique, especially in soils where the water-retention capacity is low. In Pakistan, different types of fertilizers, such as urea, DAP, MOP or SOP, are used from sowing to harvesting of crops. The use of water-retention polymers in low water-retaining soils is very important to increase its retention time. The experiment was conducted on half an acre of land in FFC Research Center located in Dyyalgarh, Millat Rd., near Deputy Wala interchange M4, Faisalabad, Punjab, Pakistan. The seed variety used in this experiment is zincole and the seed rate is 50 kg/acre. This experiment contains four different treatments, and each treatment contains three replicates. In T1, no water-retention polymers were used, while T2 was treated with standard and neem-coated urea under no water-retention polymer application. While in T3 and T4 recommended doses of polymers were used with standard and neem-coated urea. Irrigation scheduling was determined using tensiometers.. The research is aimed to keep moisture available in the root zone for better growth. For proper moisture monitoring tensiometers were installed.

Keywords: tensiometer; zincole; polymers; dry matter; neem; water retention

Citation: Anjum, L.; Rasheed, Q.; Rizwan, M.; Rehman, A.; Zamir, M.S.I. Evaluation of the Water Distribution Efficiency of Wheat under Improved Water and Fertilizer Application Techniques. *Environ. Sci. Proc.* **2022**, *23*, 25. https://doi.org/10.3390/environsciproc2022023025

Academic Editor: Muhammad Naveed Anjum

Published: 26 December 2022

1. Introduction

Agriculture imparts a vital role in the economy of Pakistan. Pakistan has the largest irrigation system in the world that contributes towards agricultural production [1]. The agriculture fulfills 40% of the world's food requirements by consuming 60% of its fresh water reserves. However, as population increases the demand for food also increases and the gap between demand and supply also widens. This gap needs to be filled by adopting new techniques for sowing crops to provide a "high-yield per drop of water" [2].

Wheat is the major crop of every agricultural country. It is very important for socio-economic growth. Wheat is required for food for both humans and animals. Wheat has a positive impact on a country's GDP. If there is shortage of wheat in a country, t will negatively affect the economic conditions as well as the population's food needs will not be met. Therefore, proper management strategies need to be adopted for the healthy growth and good yield of wheat to ensure a country's growth. Water shortage is a critical global issue, with developed countries also facing this problem. Such countries have reduced conventional farming methods and adopted new techniques to conserve soil moisture and soil water in the root zone [3]. They have applied water directly into the root zone where plants can more efficiently use this water. Certain modern techniques are known today which have been adopted by developed countries and are receiving the benefits from these techniques. The problem with modern techniques is that they require technical labor for effective use. The lack of technical labor also causes loss due to the improper use of instruments. Natural water-retention measures have also been adopted to increase water-retaining capacity of aquifers, soil, and aquatic and water-dependent ecosystems [3,4]. The agriculture sector utilizes the largest amount of water, almost 90%, and remaining water is utilized for household and industrial purposes [5]. Water availability is crucial in Pakistan because the per capita water availability is reduced to 1000 m^3 per annum, internationally considered as the water shortage threshold value [6]. In this highly stressed water profile, the major reasons which lead to the fall in the agriculture sector are population increases, rapid urbanization, and the reduction in agricultural land.

Under these circumstances, a field experiment was conducted on wheat crops using water-retention polymers in the Fouji Fertilizers Research Center of Faisalabad. The purpose of study was to develop the optimal irrigation schedules for wheat based on the different deficit levels and the following parameters were investigated, water productivity, crop yield and the development of an optimal irrigation schedule for wheat.

2. Methodology

The experimental site (geographical coordinates 31.430 N (latitude) and 73.07 degree E (longitude)) was selected at the FFC Dyyal Garh Faisalabad, Punjab, Pakistan. Wheat crop was studied from 2019–2020 to analyze the irrigation schedule and sowing techniques under RCBD statistical design (layout as shown in Table 1). Insecticide was applied to the crop as a precautionary measure to prevent the attack of insects. Fungicide was also applied to protect the crop from fungi. The use of fertilizer is very important for the better growth of any crop. For this purpose, the liquid fertilizer NPK was applied to fulfil the deficiency of macro-nutrients in the soil. There were four treatments under flood irrigation and a control under a furrow irrigation system.

Table 1. RCBD layout plan of the study area.

R3T2	R3T4	R3T1	R3T3
R2T4	R2T3	R2T2	R2T1
R1T1	R1T2	R1T3	R1T4

T1 = Recommended dose of N (standard urea), P (DAP) and K (MOP) fertilizers with four irrigations (with no application of water-retention polymers);
T2 = Recommended dose of N (neem-coated urea), P (DAP) and K (MOP) fertilizers with four irrigations (with no application of water-retention polymers);
T3 = Recommended dose of N (standard urea), P (DAP) and K (MOP) fertilizers with four irrigations (with 100% application of water-retention polymers);
T4 = Recommended dose of N (neem-coated urea), P (DAP) and K (MOP) fertilizers with four irrigations (with 100% application of water-retention polymers).

Soil and Water Sampling

The soil samples were taken from the field irrigated with groundwater flood irrigation. The water samples were collected from running tube-wells by adopting the standard procedure of sampling (Figure 1).

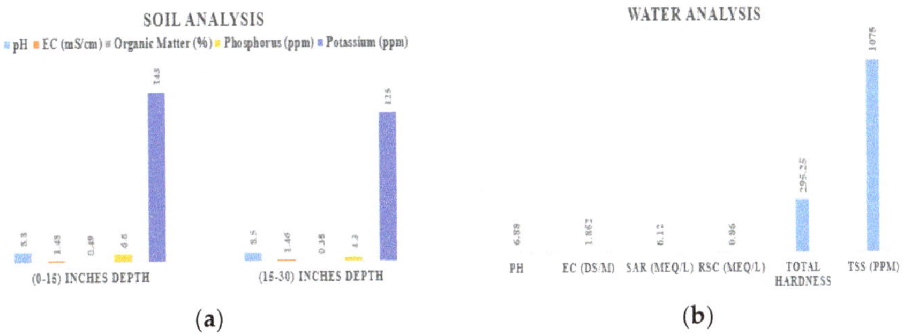

Figure 1. (**a**) Soil analysis, and (**b**) water analysis.

3. Results

Crop Harvesting Parameters

The harvesting and growth parameters are crucial to assess the crop growth over time. These are indications of the crop conditions and helps in finding the results of the experimental site. The different parameters of wheat were tested as shown in Figures 2–4.

The results shows that plant height and spike length is higher when neem-coated urea was applied. The results show that while comparing standard urea and neem-coated urea, standard urea had a good effect on crop yield. Comparing price, standard urea is cheaper than that of neem-coated urea so is a better option for crop growth in Pakistan.

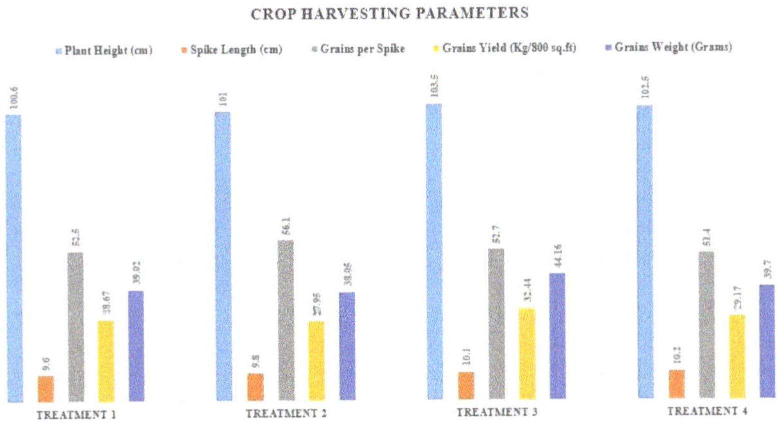

Figure 2. Graphical representation of the harvesting parameters.

Histogram and Normal Probability Plot of Crop Harvesting Parameters

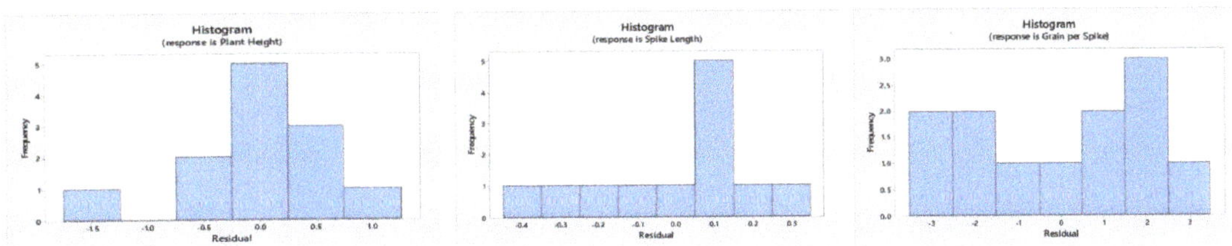

Figure 3. *Cont.*

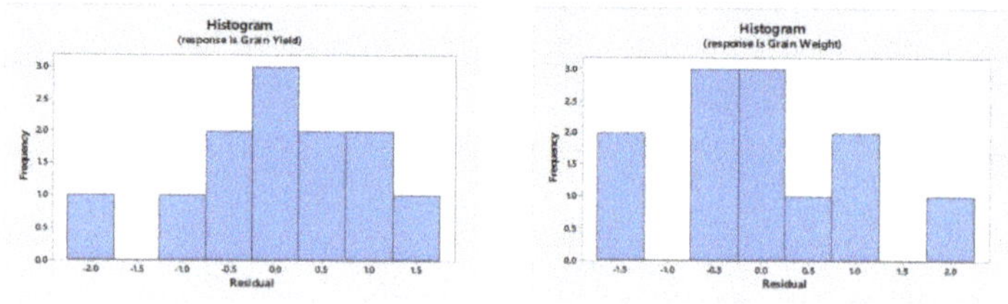

Figure 3. Histogram of the harvesting parameters.

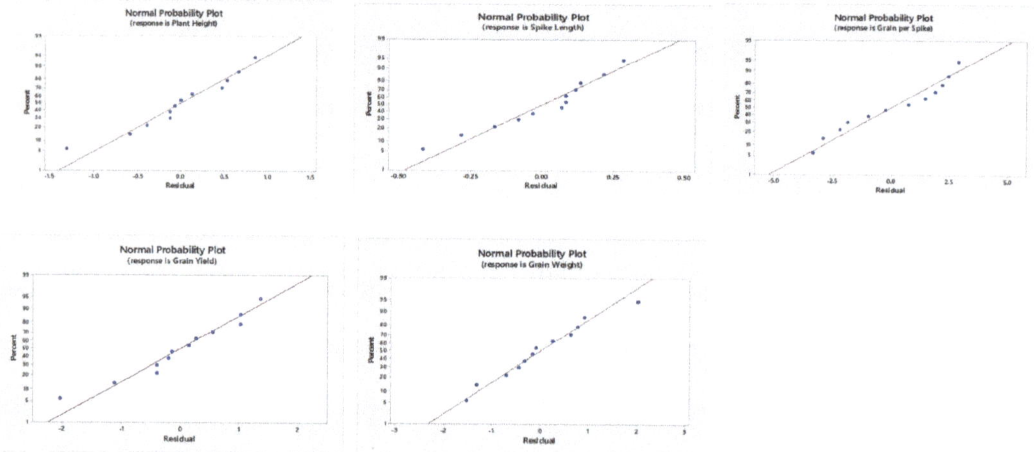

Figure 4. Normal probability plot of the harvesting parameters.

4. Conclusions

The results from the tensiometers show that water remained available to the crops when the recommended doses of the water-retention polymers were applied. The yield analysis showed that the yield was maximum with T3 treatment, where 100% of the recommended doses of the water-retention polymers with standard urea were applied. Crop conditions (plant height, grain weight, etc.) were good with T3 treatment of the recommended doses of the water-retention polymers.

Author Contributions: L.A. and Q.R. gathered and processed the data and performed the experiments. L.A. and M.R. supervised the experiments. L.A., A.R., M.R., Q.R. and M.S.I.Z. to analyzing the results and writing the manuscript. All authors have read and agreed to the published version of the manuscript.

Funding: This research received no external funding.

Institutional Review Board Statement: Not applicable.

Informed Consent Statement: Not applicable.

Data Availability Statement: The data is available upon request from the corresponding author.

Conflicts of Interest: The authors declare no conflict of interest.

References

1. Rehman, A.; Jingdong, L.; Shahzad, B.; Chandio, A.A.; Hussain, I.; Nabi, G.; Iqbal, M.S. Economic perspectives of major field crops of Pakistan: An empirical study. *Pac. Sci. Rev. B Humanit. Soc. Sci.* **2015**, *1*, 145–158. [CrossRef]
2. Iqbal, M.A.; Shen, Y.; Stricevic, R.; Pei, H.; Sun, H.; Amiri, E.; Penas, A.; del Rio, S. Evaluation of the FAO AquaCrop model for winter wheat on the North China Plain under deficit irrigation from field experiment to regional yield simulation. *Agric. Water Manag.* **2014**, *135*, 61–72. [CrossRef]
3. Anjum, L.; Ahmad, N.; Arshad, M.; Ahmad, R. Effect of different irrigation and management practices on corn growth parameters. *Pak. J. Life Soc. Sci.* **2014**, *12*, 106–113.
4. Erostate, M.; Huneau, F.; Garel, E.; Ghiotti, S.; Vystavna, Y.; Garrido, M.; Pasqualini, V. Groundwater dependent ecosystems in coastal Mediterranean regions: Characterization, challenges and management for their protection. *Water Res.* **2020**, *172*, 115461. [CrossRef] [PubMed]
5. Qureshi, R.; Ashraf, M. Water security issues of agriculture in Pakistan. *PAS Islamabad Pak.* **2019**, *1*, 41.
6. Nabi, R.B.S.; Tayade, R.; Hussain, A.; Kulkarni, K.P.; Imran, Q.M.; Mun, B.-G.; Yun, B.-W. Nitric oxide regulates plant responses to drought, salinity, and heavy metal stress. *Environ. Exp. Bot.* **2019**, *161*, 120–133. [CrossRef]

environmental sciences proceedings

MDPI

Proceeding Paper

Cost–Benefit Analysis of Solar Photovoltaic Energy System in Agriculture Sector of Quetta, Pakistan †

Zaid Mustafa [1,*], Rashid Iqbal [1], Mahwish Siraj [1] and Iqbal Hussain [2]

1 Department of Agricultural Sciences, Allama Iqbal Open University, Islamabad 44000, Pakistan
2 Department of Forestry, Environment and Wildlife, Government of Khyber Pakhtunkhwa, Peshawar 24550, Pakistan
* Correspondence: zaid.mustafa9@gmail.com
† Presented at the 1st International Precision Agriculture Pakistan Conference 2022 (PAPC 2022)—Change the Culture of Agriculture, Rawalpindi, Pakistan, 22–24 September 2022.

Abstract: The energy crisis in Pakistan has amplified the need for solar photovoltaic (PV) technologies in the agriculture sector. Currently, solar PV systems in Pakistan are primarily used for water-pumping irrigation. This article presents an investigation of the cost–benefit analysis of solar photovoltaic energy systems in the agriculture sector in the Baluchistan province of Pakistan. The findings of the study reveal that solar PV systems are relatively economical, as a benefit-to-cost ratio for the solar system is calculated to be 9.3 as compared to grid electricity which is calculated to be 8.4. Furthermore, solar photovoltaics can increase agricultural productivity substantially by providing a continuous power supply for water-pumping irrigation. However, the high initial cost and weather dependency of solar systems are the main obstacles to adopting PV technologies in the agriculture sector. Nevertheless, inconsistent grid power supply and sky-rocketing energy costs in Pakistan cause the local farmers to shift to solar PV systems for water-pumping irrigation to boost their agricultural productivity.

Keywords: renewable energy; cost–benefit analysis; cost-to-benefit ratio

Citation: Mustafa, Z.; Iqbal, R.; Siraj, M.; Hussain, I. Cost–Benefit Analysis of Solar Photovoltaic Energy System in Agriculture Sector of Quetta, Pakistan. *Environ. Sci. Proc.* **2022**, *23*, 26. https://doi.org/10.3390/environsciproc2022023026

Academic Editor: Muhamamd Umair

Published: 27 December 2022

1. Introduction

The energy crisis in Pakistan not only affects the domestic life of the people but also hinders the economic development of the country [1]. Electricity shortages impose a high cost on the economy as a whole, which is estimated to be approximately two percent of the country's annual GDP [2]. The long power outage has detrimental effects on every sector that consumes electricity including domestic, commercial, industrial, and agricultural sectors [3].

Although Pakistan has abundant renewable energy resources which are more than sufficient to meet the present and future electricity demands in various sectors including agriculture [4], the current share of renewable energy is insufficient for the total energy mix in the country [5]. However, the local people, particularly in rural areas of Pakistan, are shifting to renewable energy, mainly solar PVs [4], as solar energy is richly available in most parts of the country and has the potential to be effectively utilized for electricity generation [6] to fill the demand–supply gap in the agriculture sector [1].

The Balochistan province of Pakistan has the highest average sunshine hours in the world [7], which provides a viable choice for installing standalone solar PVs in remote arable areas for groundwater harvesting. However, at present, solar PV systems are occupying a central position in the agricultural sector of the province [8] as, in Balochistan, solar panels are an agreeable option being economically viable [9]. The present study is an investigation and comparison of a cost–benefit analysis of solar PV systems and grid electricity use in the agriculture sector in the current practicing scenario.

2. Materials and Methods

This article employs the cost–benefit analysis (CBA) technique to ascertain the economic benefits and cost-effectiveness of solar photovoltaic technologies in the agriculture sector. Cost–benefit analysis is a process in which economists sum the benefits and then subtract them from the total cost [10]. The benefit–cost ratio was calculated separately for both grid electricity and photovoltaic solar system to determine the most economical alternative.

Benefit–cost ratio B/C is calculated by the formula

$$\frac{P_{VB}}{P_{VC}} = \frac{F_i\left[\frac{(1+d)^n-1}{d(1+d)^n}\right]}{C_o} \tag{1}$$

where P_{VB} denotes present value of benefit, and P_{VC} denotes present value of cost, while F_i, C_o, d, and n represent cash inflow, cash outflow, discount rate, and number of years, respectively.

The study was carried out in the Balochistan province of Pakistan, which receives an average daily global irradiation of about 19–20 MJ/m^2 and average daily sunshine of about 8–8.5 h [11]. For this study, 392 farmers using solar PV technologies for water pumping were randomly interviewed. Furthermore, data pertaining to grid electricity were obtained from Quetta Electric Supply Company (QESCO). The comparison between the solar photovoltaic system and grid electricity system was meant to probe into the relatively economical alternative for the agriculture sector in the study area. Moreover, costs/prices of photovoltaic solar panels and other accessories were retrieved from MRS-2020 and apple crop was treated as an assumed crop subject to the anticipated increase in yield.

3. Results and Discussion

3.1. Annual Power Costs and Production Differences

The annual power cost and production differences are shown in the following figures (Figures 1 and 2).

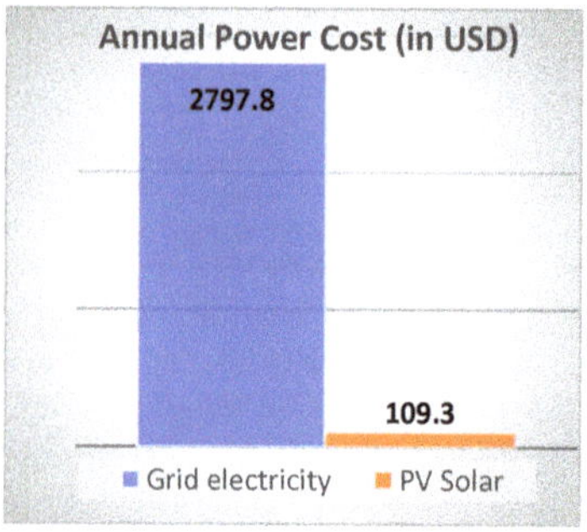

Figure 1. Annual power cost.

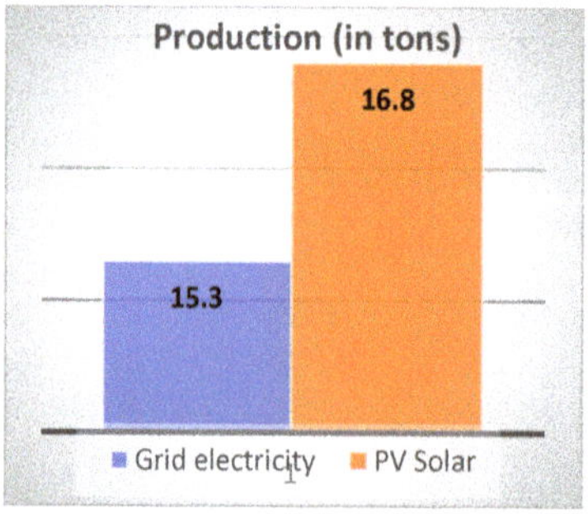

Figure 2. Production.

3.2. Cost–Benefit Analysis

Cost–benefit analysis was applied to ascertain the net benefit of installing a solar photovoltaic system in terms of savings in expenditures and increase in agricultural yield as a result of consistent power supply for water pumping. For this purpose, all related costs and benefits were calculated and compared to decide upon the cost-effectiveness and ensuing benefits of installing a PV solar system as an alternative to grid electricity. Table 1 shows all the costs associated with a water-pumping system run by grid electricity and required for an average of a 5.7-acre piece of land.

Table 1. Costs related to grid electricity.

Cost Details	Particulars	Cost (in USD)
Initial cost	HT (structure) 46′	619.4
	LT (structure) 30′8″	162.3
	25 KVA11/4 KV transformer	859.7
	Static energy meter 3 phase	19.2
	AAC ANT (1/172)	298.4
	11kv steel cross arm	25.6
	11 kv D/Out C/Out insulator	24.7
Connection cable	P.V.C 7/058 I/Core (10 mm^2)	311.0
Submersible	20 hp	1220.0
Pipe	40 (20 ft each)	1376.5
Carriage	Not applicable	21.6
Annual O and M	12 months	2797.8
Other expenses	Not applicable	109.2
Total		**7845.4**

Table 2 indicates all the associated costs of installing a PV solar system derived from an analysis of primary data. The rates are based on the average rates of 392 respondents using a PV solar system for water-pumping irrigation in the study area.

Table 2. Costs related to PV solar system.

Cost Details	Particulars	Costs (in USD)
Initial cost	Solar panels 300 W (65)	6134.0
	DC inverter	959.1
	Connection cable	83.8
	Frames	1629.3
Submersible	20 hp	1220.0
Pipe	40 (20 ft each)	1376.5
Carriage	Not applicable	21.6
Annual O and M	12 months	26.6
Other expenses	Not applicable	109.2
Total		**11,560.3**

According to the Government of Balochistan, the annual per-acre production of apples is 2.676 tons in Balochistan, approximately. Assuming the same production for a respondent growing an apple orchard in agricultural fields in the study area, the total apple yield is calculated to be 15.252 tons for an area of 5.7 acres when all the other factors are assumed to be constant. An analysis of the primary data indicates an average increase of 9.9% in agricultural production as a result of using a PV solar system for water pumping, increasing the total yield for an area of 5.7 acres from 15.252 tons to 16.763 tons per annum.

$$= 15.252 + (15.252 \times 9.9\%)$$
$$= 16.783 \text{ tons}$$

According to Agriculture Marketing Information Service 2020, the price of apples per ton is USD 553.4, so the total amount for 15.252 tons is calculated to be USD 8502.9, and for 16.763 is calculated to be USD 9276.7, with a net difference of USD 773.8 per annum.

Based on the assumptions and analysis of primary data, Benefit–Cost Ratio (CBR) was applied separately for grid electricity and PV solar. The discount rate is taken as 7% for 25 years of life span of PV solar system.

3.2.1. Benefit–Cost Ratio for Grid Electricity

Total initial investment (C_o) = USD 7845.4
Annual cash flow (F_i) = USD 5669.2 (total income—O and M)
Life span (n) = 25 years
Discount rate (d) = 7%
By putting the values in the formula, we get

$$\frac{P_{VB}}{P_{VC}} = 8.4$$

3.2.2. Benefit–Cost Ratio for PV Solar System

Total initial investment (C_o) = USD 11560.3
Annual cash flow (F_i) = USD 9250.1 (total income—O and M)
Life span (n) = 25 years
Discount rate (d) = 7%
By putting the values in the formula, we get

$$\frac{P_{VB}}{P_{VC}} = 9.3$$

The results reveal that both the alternatives are feasible as $8.4 > 1$ and $9.3 > 1$; however, the benefit–cost ratio suggests that a PV solar system is relatively economical compared with grid electricity for water-pumping irrigation in the agriculture sector.

4. Conclusions and Recommendations

An ocularly photovoltaic solar system seems to be the economical alternative to grid electricity for water pumping in the agriculture sector. However, the results of the study bring to the surface a slight difference between the two in terms of benefits. The findings discern that BCR for solar photovoltaic systems differs from grid electricity by just 0.9 which needs further improvement to raise its efficiency primarily through multi-junction PV cells. Nevertheless, the crop productivity can be enhanced substantially through consistent power supply for water pumping in areas where grid electricity is hard to reach.

Author Contributions: R.I. conducted the research under the supervision of Z.M.; M.S. and I.H. critically reviewed and improved the manuscript linguistically. All authors have read and agreed to the published version of the manuscript.

Funding: This research received no external funding.

Institutional Review Board Statement: Not applicable.

Informed Consent Statement: Not applicable.

Data Availability Statement: Not applicable.

Acknowledgments: The authors are grateful to their colleagues at the Department of Agricultural Sciences, AIOU, and Islamabad for their valuable comments and suggestions to the original version of this article.

Conflicts of Interest: All authors declare that there are no competing interest.

References

1. Irfan, M.; Zhao, Z.Y.; Ahmad, M.; Mukeshimana, M.C. Solar energy development in Pakistan: Barriers and policy recommendations. *Sustainability* **2019**, *11*, 1206. [CrossRef]
2. Abbasi, K.R.; Abbas, J.; Tufail, M. Revisiting electricity consumption, price, and real GDP: A modified sectoral level analysis from Pakistan. *Energy Policy* **2021**, *149*, 112087. [CrossRef]
3. Yaqoob, H.; Teoh, Y.H.; Din, Z.U.; Sabah, N.U.; Jamil, M.A.; Mujtaba, M.; Abid, A. The potential of sustainable biogas production from biomass waste for power generation in Pakistan. *J. Clean. Prod.* **2021**, *307*, 127250. [CrossRef]
4. Shah, S.A.A.; Solangi, Y.A. A sustainable solution for electricity crisis in Pakistan: Opportunities, barriers, and policy implications for 100% renewable energy. *Environ. Sci. Pollut. Res.* **2019**, *26*, 29687–29703. [CrossRef] [PubMed]
5. Wakeel, M.; Chen, B.; Jahangir, S. Overview of energy portfolio in Pakistan. *Energy Procedia* **2016**, *88*, 71–75. [CrossRef]
6. Solangi, Y.A.; Shah, S.A.A.; Zameer, H.; Ikram, M.; Saracoglu, B.O. Assessing the solar PV power project site selection in Pakistan: Based on AHP-fuzzy VIKOR approach. *Environ. Sci. Pollut. Res.* **2019**, *26*, 30286–30302. [CrossRef] [PubMed]
7. Shah, S.A.A.; Valasai, G.D.; Memon, A.A.; Laghari, A.N.; Jalbani, N.B.; Strait, J.L. Techno-economic analysis of solar pv electricity supply to rural areas of Balochistan, Pakistan. *Energies* **2018**, *11*, 1777. [CrossRef]
8. Ali, M.A.; Yamin, M.; Arslan, C.; Farid, U.; Nasir, A. A Feasibility Study: Off-Grid Photovoltaic Solar Power Supply to the Remote Areas of Pakistan. *Pak. J. Agri. Sci.* **2020**, *57*, 1313–1316.
9. Kakar, M.Z.K.; Harijan, K.; Kumar, L. Load Assessment of Solar PV System: Case Study of Tor Tang Village of Baluchistan. *Int. J. Renew. Energy Resour.* **2018**, *8*, 29–35.
10. O'Mahony, T. Cost-Benefit Analysis and the environment: The time horizon is of the essence. *Environ. Impact Assess. Rev.* **2021**, *89*, 106587. [CrossRef]
11. Waheed, A. To Electrify a Flood Affected Village in Sajawal, Pakistan by Utilizing the Available Renewable Energy Resources. Available online: http://urn.kb.se/resolve?urn=urn:nbn:se:kth:diva-149865 (accessed on 20 September 2022).

environmental sciences
proceedings

MDPI

Proceeding Paper

Potential Application of Soil Probiotics for Sustainable Soil Health and Improved Peanut Yield †

Nabeeha Javed [1,2,*], Shahzada Sohail Ijaz [2], Qaiser Hussain [1,2], Rabia Khalid [2], Shoaib Rashid Saleem [1,3,4], Sehrish Kanwal [1,5], Muhammad Naveed Tahir [1,6] and Basit Shahzad [1,7]

1 Data-Driven Smart Decision Platform, PMAS Arid Agriculture University, Rawalpindi 46000, Pakistan
2 Institute of Soil & Environmental Sciences, PMAS Arid Agricultural University, Rawalpindi 46000, Pakistan
3 Center for Precision Agriculture, PMAS Arid Agricultural University, Rawalpindi 46000, Pakistan
4 Department of Farm Machinery and Precision Engineering, PMAS Arid Agricultural University, Rawalpindi 46000, Pakistan
5 Department of Entomology, PMAS Arid Agriculture University, Rawalpindi 46000, Pakistan
6 Department of Agronomy, PMAS Arid Agriculture University, Rawalpindi 46000, Pakistan
7 Department of Horticulture, PMAS Arid Agricultural University, Rawalpindi 46000, Pakistan
* Correspondence: jnabeeha@yahoo.com
† Presented at the 1st International Precision Agriculture Pakistan Conference 2022 (PAPC 2022)—Change the Culture of Agriculture, Rawalpindi, Pakistan, 22–24 September 2022.

Abstract: Conventional agricultural practices and a rapidly growing population have both contributed to an increase in interest in cutting-edge research on environmental friendly farming methods. A field experiment to investigate the potential of soil probiotics on soil and plant health is under process at IOT Smart Research Farm, PMAS-Arid Agriculture University Rawalpindi from 22 June till mid October 2022. The plots contain four different treatments (Control, Full dose NPK, Probiotics only and $\frac{1}{2}$ NPK+ Probiotics) on peanut crop. Probiotics (*Actinomycetes* sp. & *Mycobacterium neoaurum*) were applied through seed coating. Treatments were triplicated in a randomized complete block design. Different plant physiological characteristics (height, canopy, no of leaves, leaf area index, chlorophyll content, and normalized difference vegetation index) and soil properties (pH, Ec, moisture, nitrogen, phosphorous, and potassium) are under investigation. For different plant parameters different novel devices are being used, such as leaf area index meters to find the index area, and chlorophyll meters for chlorophyll content, whereas for soil parameters proximal sensors are being used. The findings of the trial up till now show the best results in $\frac{1}{2}$ NPK+ probiotics followed by probiotics, full-dose NPK and control which are encouraging, indicating enhanced crop productivity and improved soil health. This study will provide a way out for increased peanut production in an environmental friendly manner for farmers.

Keywords: NPK; NDVI; LAI

Citation: Javed, N.; Ijaz, S.S.; Hussain, Q.; Khalid, R.; Saleem, S.R.; Kanwal, S.; Tahir, M.N.; Shahzad, B. Potential Application of Soil Probiotics for Sustainable Soil Health and Improved Peanut Yield. *Environ. Sci. Proc.* **2022**, *23*, 27. https://doi.org/10.3390/environsciproc2022023027

Academic Editor: Muhammad Naveed Anjum

Published: 27 December 2022

1. Introduction

The peanut (*Arachis hypogaea* L.), also known as groundnut, is a member of the family Leguminosae, which also includes plants that produce subterranean fruits commonly called shell beans [1]. Planting peanuts improves soil quality because of the legume's ability to draw atmospheric nitrogen and convert it into usable forms. Because peanuts can utilize nitrogen from the air, they need only trace amounts of N to thrive [2].

Conventional agricultural practices and a rapidly growing population have both contributed to an increased interest in cutting-edge research on environmentally friendly farming methods. The incorporation of plant probiotics is one method for achieving the fundamental objectives of sustainable agriculture, which include the maintenance of a healthy environment, social and economic fairness, and the sustainability of the agricultural economy as population growth and food demand are inextricably linked) [3].

As the global population continues to grow, there is a need to increase food production to meet population's food needs. Furthermore, arable land has lost its natural nutrients potential over time. As a result, alternative methods such as fertilizers, insecticides, and herbicides are now utilized to reinforce the soil and increase its output [4].

This approach will not only accomplish the primary objectives of sustainable agriculture, but it will also increase the variety of microorganisms found in the soil. Increased photosynthesis and the production of bioactive substances such as plant growth regulators and enzymes, disease control, accelerated decomposition of lignin materials, stimulate the decomposition of organic wastes and residues, and release inorganic nutrients for plant uptake are just some of the ways in which the use of probiotics in agriculture can boost crop growth and yield [5].

This research is conducted with the following objectives;

- Evaluate the potential of selected microbes under natural field conditions on soil health;
- Identify their beneficial effects on crop productivity.

2. Methodology

A field experiment is under process at the IOT (internet and other things) Smart Research Farm, Arid Agriculture University Research Farm, Chakwal Road, Koont from 22 June till mid October 2022. The plots contain four different treatments (control, full-dose NPK, probiotics only and $\frac{1}{2}$ NPK + probiotics) on peanut crop. Probiotics (*Actinomycetes* sp. and *Mycobacterium neoaurum*) were applied through seed coating. Treatments were triplicated in a randomized complete block design. For irrigation purposes rainwater is used. Different plant physiological characteristics (height, canopy, no of leaves, leaf area index, chlorophyll content, and normalized difference vegetation index) and soil properties (pH, EC, moisture contents, nitrogen, phosphorous, and potassium) are under investigation. For different plant parameters different novel devices are being used, such as leaf area index meters to find the index area, and chlorophyll meters for chlorophyll content, whereas for soil parameters proximal sensors are being used.

The preliminary data collected for various characteristics was subjected to analysis of variance (ANOVA) and means compared at a 5% level of significance by least significance difference (LSD) tests [6].

3. Results

3.1. Experimental Soil Properties

The results of soil pH of the experimental area (Figure 1) shows a slightly acidic pH of $\frac{1}{2}$NPK + probiotics, whereas the pH of the control is more acidic. Peanuts grows better in slightly acidic soil (pH 6–6.5). The highest pH was observed in soil probiotics (pH 7.88) whereas the lowest pH was observed in the control (6.79). The moisture content of the field was greater in the control compared to other treatments, the control shows 24% soil water content, whereas traditional practice NPK application shows a lower water content of 15%. The pH and moisture content are shown in Figure 2.

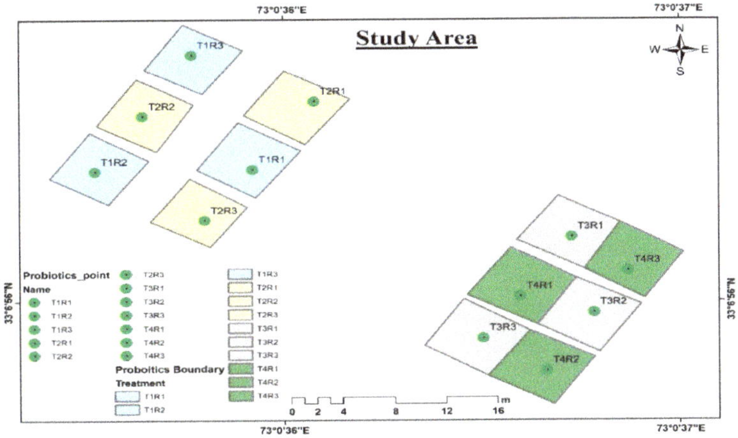

Figure 1. Study map of the experimental site.

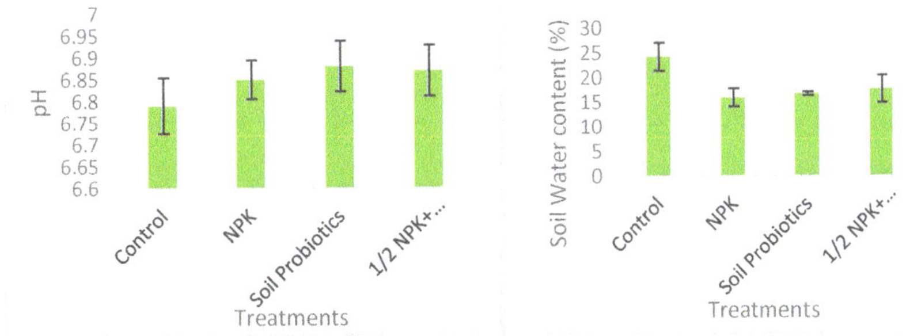

Figure 2. pH and moisture of the experimental site.

Leaf Area Index of Peanut Crop

Soil probiotics positively influenced the leaf area index (LAI) of the crop. The LAI of the $\frac{1}{2}$ NPK + probiotics treatment was greatest among all treatments, whereas the control showed the lowest LAI. The trend in LAI observed was $\frac{1}{2}$ NPK + probiotics > probiotics > NPK > control (Figure 3). At initial stages of the experiments show encouraging results.

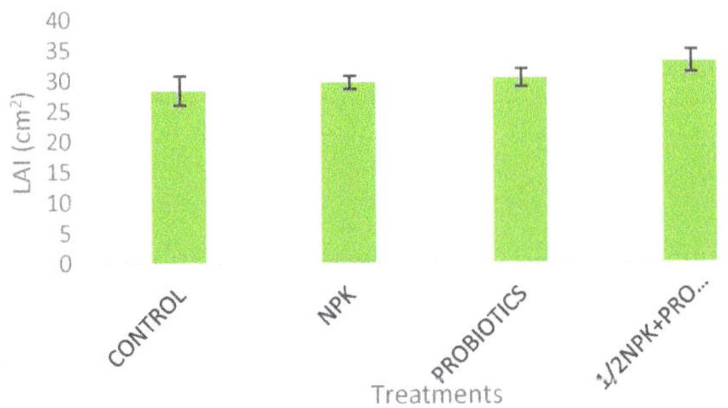

Figure 3. Leaf area index of the peanut plants.

4. Conclusions

Biofertilizers are a safe farm product for consumers and have been shown to increase food production; as a result, they continue to be the best option for ensuring the safety of crops and increasing global food security. The shortage of plant nutrients relative to their supply due to chemical fertiliser has been greater than 10 million tonnes in recent years. The initial investment and ongoing operating expenses of fertiliser facilities make long-term reliance on these inputs which is an unsustainable strategy, both financially and in terms of their impact on the environment. To combat this, biofertilizers and other forms of sustainable fertilisers are a need of hour. Therefore, probiotics, another form of biofertilizers, is an alternative approach to supress disease and pest attacks, enhance the NPK content of soil and increase nodulation that can improve the soil nitrogen contents.

Author Contributions: Conceptualization, N.J. and Q.H.; methodology, N.J., S.S.I. and R.K., formal analysis, N.J., B.S. and S.K.; investigation, S.R.S.; resources, Q.H.; data curation, N.J.; writing—original draft preparation, N.J., Q.H. and R.K.; writing—review and editing, N.J.; visualization, Q.H.; supervision, Q.H.; project administration, M.N.T.; funding acquisition, N.J. All authors have read and agreed to the published version of the manuscript.

Funding: The study is part of a pilot project for the Data-Driven Smart Decision Platform for Increased Agriculture Productivity (DDSDP) PSDP-funded project No. 332".

Institutional Review Board Statement: Not applicable.

Informed Consent Statement: Not applicable.

Data Availability Statement: Not applicable.

Acknowledgments: The authors would like to thank IS, ES and DDSDP project.

Conflicts of Interest: The authors declare no conflict of interest.

References

1. Aboelill, A.A.; Mehanna, H.M.; Kassab, O.M.; Abdallah, E.F. The Response of Peanut to Foliar Spraying with Potassium Under Water Stress Conditions. *Aust. J. Basic Appl. Sci.* **2012**, *6*, 626–634.
2. Jordan, D.L.; Brandenburg, R.; Brown, B.; Bullen, G.; Roberson, G.; Shew, B. *Peanut Information*; AG-331; North Carolina Cooperative Extension Service Publications: Raleigh, NC, USA, 2017; p. 172.
3. Jadon, K.S.; Thirumalaisamy, P.P.; Koradia, V.G.; Padavi, R.D. Management of peanut (*Arachis hypogaea* L.) diseases through nutrient supplements. *Legume Res. Int. J.* **2018**, *41*, 316–321. [CrossRef]
4. Xie, J.; Shi, H.; Du, Z.; Wang, T.; Liu, X.; Chen, S. Comparative genomic and functional analysis reveal conservation of plant growth promoting traits in Paenibacillus polymyxa and its closely related species. *Sci. Rep.* **2016**, *6*, 21329. [CrossRef] [PubMed]
5. Chen, Y.N.; Wei, W.H.; Ren, X.P.; Zhao, X.Y.; Zhou, X.J.; Huang, L.; Tang, X.C.; Jiang, H.F. Construction of a high-quality genomic BAC library for Chinese peanut cultivar Zhonghua 8 with high oil content. *Bot. Stud.* **2014**, *55*, 8. [CrossRef] [PubMed]
6. Steel, R.G.; Torrie, J.H. *Principles and Procedures of Statistics: A Biometrical Approach*; McGraw-Hill Book Company: New York, NY, USA, 1986; pp. 401–437.

environmental sciences proceedings

MDPI

Proceeding Paper

The Role of Sustainable Land and Water Conservation Practices in Flood Mitigation †

Muhammad Sohail Waqas [1,2,*], Muhammad Jehanzeb Masud Cheema [3,4], Saddam Hussain [5,6], Muhammad Mazhar Iqbal [7], Lubna Anjum [5] and Muhammad Abubakar Aslam [4]

1 Soil Conservation Group, Agriculture Department (Field Wing), Government of the Punjab, Rawalpindi 46000, Pakistan
2 Agricultural and Biological Engineering, Tropical Research and Education Center, Institute of Food and Agricultural Sciences, University of Florida, Homestead, FL 33031, USA
3 Faculty of Agricultural Engineering and Technology, PMAS-Arid Agriculture University, Rawalpindi 46000, Pakistan
4 National Center for Industrial Biotechnology, PMAS-Arid Agriculture University, Rawalpindi 46000, Pakistan
5 Department of Irrigation and Drainage, University of Agriculture Faisalabad, Faisalabad 38000, Pakistan
6 Department of Agricultural and Biological Engineering, University of California (UC Davis), Davis, CA 95616, USA
7 Water Management Training Institute (WMTI), Department of Agriculture (Water Management Wing), Government of the Punjab, Lahore 54000, Pakistan
* Correspondence: sohailwaqasrana@gmail.com

† Presented at the 1st International Precision Agriculture Pakistan Conference 2022 (PAPC 2022)—Change the Culture of Agriculture, Rawalpindi, Pakistan, 22–24 September 2022.

Abstract: Promising Land and Water Conservation Practices (LWCPs) play a vital role in restraining floods and keeping the land productive. Floods are usually destructive and increase the risk of drowning, waterborne diseases, malnutrition, and multiple long-term knock-on effects. Similarly, the 2022 monsoon season has prompted the most severe flood in Pakistan. The main objective of this review is to highlight the importance of LWCPs as an adaptation strategy for flood mitigation. Moreover, different LWCPs are discussed concerning studies carried out in different regions and published in scientific journals, technical reports, and notes from experts. It was observed that both the in situ and ex situ LWCPs have a significant effect on reducing land degradation and flood control. Additionally, most of the reviewed studies showed a positive impact of LWCPs on agricultural productivity, primarily due to the retention of nutrients and moisture. Hence, land conservation practices including biological and agronomic measures (i.e., contour farming, conservation tillage, strip cropping, vegetation, etc.), and mechanical or engineering methods (i.e., check dams, bunding, ponds/reservoirs, etc.) are aimed at reducing the run-off velocity and mitigating the floods. The results of this study will encourage the stakeholders to adopt LWCPs to lessen flood hazards and uplift agricultural production by limiting the land degradation processes.

Keywords: soil conservation; runoff; flood risk management; check dams; agronomic measures; watershed management

Citation: Waqas, M.S.; Cheema, M.J.M.; Hussain, S.; Iqbal, M.M.; Anjum, L.; Aslam, M.A. The Role of Sustainable Land and Water Conservation Practices in Flood Mitigation. *Environ. Sci. Proc.* **2022**, *23*, 28. https://doi.org/10.3390/environsciproc2022023028

Academic Editor: Tahir Iqbal

Published: 29 December 2022

1. Introduction

Flooding is the most common natural disaster and can cause widespread devastation, resulting in loss of life and damage to infrastructure, and it can compromise medium-to-long-term health impacts. According to WHO (2022) [1] more than two billion people worldwide were affected by floods from 1998 to 2017, worldwide. Similarly, a devastating flood hit Pakistan during the 2022 monsoon rains and inundated thousands of hectares of agricultural land, drowned livestock, damaged infrastructure, and wiped out urban areas. The damage was particularly severe in southern Punjab and the Sindh and Balochistan provinces (Figure 1). Heavy and prolonged rainfalls, snow melts, and unmanaged watersheds are the main drivers of destructive floods. Two main factors regulate the flood, i.e., runoff depth and peak discharge [2]. Watershed management through promising LWCPs reduce runoff and sediment; this reduction is generally achieved by altering the runoff processes. The high-intensity runoff erodes the topsoil and contributes to the sedimentation of downstream reservoirs, consequently decreasing the live water storage of the reservoirs [3].

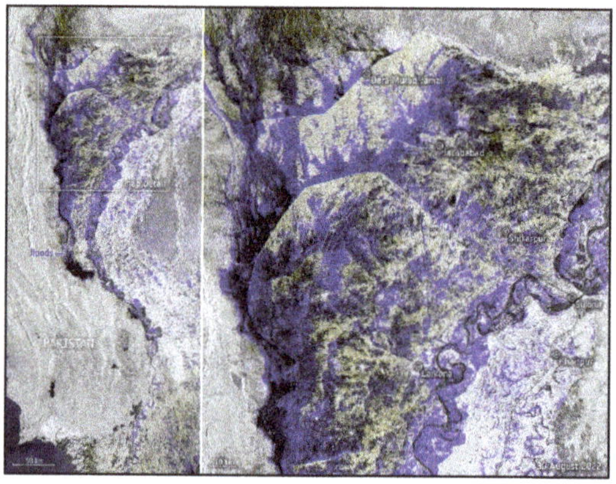

Figure 1. Detailed flood inundation map of southern Punjab, Sindh, and Balochistan (The image contains modified Copernicus Sentinel data (2022), processed by ESA, CC BY-SA 3.0 IGO—*Source:* [4]).

Soil erosion has been a critical issue and has induced numerous inflated environmental issues. Yousuf and Singh, (2019) [5] stated that different degradation mechanisms have degraded almost 52% of the total productive land. Moreover, due to the acute shortage of water [6–8], and under the emerging climate change scenarios [9], we must emphasize the expansion and adoption of effective measures for land and water conservation to mitigate floods.

Thus, mitigation against floods could be achieved by reducing the runoff intensity and by increasing the runoff lag time through standard LWCPs and flood protection measures. These practices include contour farming, conservation tillage, strip cropping, crop rotation, cover crops, water diversion channels, spillways (gully control), field outlets, terracing, spurs, check dams, and rainwater harvesting structures.

2. Land and Water Conservation Practices

LWCPs help to conserve water, reduce erosion, keep the land fertile, reduce runoff intensity, and mitigate devastating floods. Following are the two broader categories of LWCPs that can sustain the environment and enhance agricultural production.

2.1. *Agronomical or Biological Practices*

These practices are usually in-situ and reduce the impact of raindrops by protecting the soil surface, and improving the infiltration rate, resulting in less runoff generation and soil loss due to erosion. Vegetation or cover crops are one of the agronomic practices to tackle soil erosion and reduce the runoff intensity. All of the agricultural operations starting from plowing, making ridges and furrows, planting, inter-cultural, etc. are performed along contours, called contour farming. These operations hinder the runoff and reduce its velocity, which in turn reduces soil loss. Moreover, the runoff intensity also depends on the selection of crops, the substantial biomass, the widespread rooting system, and the dense canopy cover; these parameters reduce the erosive impact of rainfall and hinder the runoff [10]. The close-growing and dense cover crops are the best for controlling the erosion process, i.e., groundnut, cowpea, grams, etc. Similarly, other agronomical or biological practices are listed in Table 1.

Table 1. Some findings from the literature regarding land and water conservation practices.

LWCPs	Main Findings	Reference
Agronomical/Biological measures		
Grassed waterway	Decreases runoff coefficient and reduces nutrients loss	[11]
Conservation tillage	Reduces runoff, improves infiltration rate, and reduces evaporation loss	[5]
Strip cropping	Reduces the runoff velocity, increases the time of concentration, conserves soil moisture, and increases crop production	[5]
Land configuration methods (i.e., bed- and ridge-furrow)	Reduces runoff and soil and nutrient loss, conserves soil moisture, and higher gain in productivity and profitability	[5,10]
Agroforestry measures	Reduces soil erosion by up to 10%, reduces runoff, and increases water infiltration	[10]
Mechanical/Engineering methods		
Terracing	Reduces runoff velocity, improves infiltration, and controls soil erosion	[5,12]
Field outlet	Safe disposal of rainwater from one field to another	—
Water diversion channel	Divert runoff water and control runoff losses	[10]
Spillway	Disposes of runoff water safely and reclaims severely eroded gullies	-
Bunding (i.e., contour bunding, graded bunding, and peripheral bunds)	Conserves soil moisture, disposes of excess runoff water safely, and reduces erosion	[10]
Contour trenching	Reduces runoff velocity and conserves soil moisture	[5,10]

2.2. Mechanical or Engineering Methods

Mechanical measures or engineering methods of soil and water conservation are mainly designed to alter the land slope, lessen the runoff velocity and sedimentation, and safely transport the runoff water from the upstream to the downstream side. Small rainwater harvesting ponds and dams are suitable for plugging gullies and controlling erosion and thus reducing runoff velocity. Similarly, check dams are also built across gullies to confront the velocity of intense runoff. Moreover, preventing a flood from turning into a calamity is a fundamental goal of a check dam system. Wang et al. (2021) [13] also reported that check dams significantly increase runoff infiltration and effectively mitigate flood processes. The importance of other mechanical/engineering methods is listed in Table 1.

3. Discussion

The LWCPs play a significant role in controlling erosive runoff by reducing the flow velocity; moreover, these practices enhance the time of concentration, thus increasing the infiltration and crop production. Moreover, strip copping had an effective role in regulating surface runoff and floods. Check dams are one of the prominent measures of mechanical/engineering methods, reducing the flood peak and flood volume and thus mitigating the floods [5].

Globally, check dams are used to control soil and water erosion, reduce runoff velocity, and improve infiltration. Wang et al. (2021) [13] reported that runoff lag times significantly increased by constructing check dams. Similarly, the construction of small dams also obstructs the runoff and discourages the flood processes [5]. Hence, LWCPs significantly reduce soil erosion, trap silt, reduce runoff velocity by hindering flowing water, and evade the risk of floods by changing local geomorphology, thus maintaining environmental sustainability.

4. Conclusions

LWCPs are imperative in controlling land degradation, reducing runoff velocity, and thus mitigating the risk of devastating floods. Moreover, these practices increase infiltration and in situ moisture conservation, improve soil and water quality, and in turn effectively enhance overall farm production. Therefore, the holistic adoption of these practices at watershed levels is crucial for flood mitigation, ecosystem protection, and sustaining agricultural productivity. Moreover, new policies regarding regional land and water conservation must be framed, keeping in view the damages perceived from the recent floods, and the post-assessment of these policies must be followed to increase efficiency.

Author Contributions: Conceptualization, M.S.W. and M.J.M.C.; methodology, M.S.W., M.J.M.C., and S.H.; investigation, M.S.W. and S.H.; resources, M.S.W., S.H., M.M.I., L.A., and M.A.A.; data curation, M.S.W., S.H., M.M.I., and M.A.A.; writing—original draft preparation, M.S.W.; writing—review and editing, M.S.W., M.J.M.C., and S.H.; visualization, M.S.W.; supervision, M.J.M.C. All authors have read and agreed to the published version of the manuscript.

Funding: This research received no external funding.

Institutional Review Board Statement: Not applicable.

Informed Consent Statement: Not applicable.

Data Availability Statement: Not applicable.

Conflicts of Interest: The authors declare no conflict of interest.

References

1. WHO. Available online: https://www.who.int/health-topics/floods#tab=tab_2 (accessed on 2 October 2022).
2. Yuan, S.; Li, Z.; Li, P.; Xu, G.; Gao, H.; Xiao, L.; Wang, F.; Wang, T. Influence of check dams on flood and erosion dynamic processes of a small watershed in the Loss Plateau. *Water* **2019**, *11*, 834. [CrossRef]
3. Amira, R.M.; Ali, S.; Cheema, M.J.M.; Hussain, S.; Waqas, M.S.; Arshad, R.H.; Salam, M.; Raza, A.; Aslam, M. Estimating Sediment Yield At Tarbela Dam And Flood Forecasting Through Continuous Precipitation-Runoff Modeling Of Upper Indus Basin. *Big Data Water Resour. Eng. (BDWRE)* **2020**, *1*, 43–48. [CrossRef]
4. European Space Agency. Available online: https://www.esa.int/ESA_Multimedia/Images/2022/09/Pakistan_inundated (accessed on 2 October 2022).
5. Yousuf, A.; Singh, M. *Watershed Hydrology, Management and Modeling*; CRC Press: Boca Raton, FL, USA, 2019; pp. 77–97.
6. Iqbal, M.M.; Li, L.; Hussain, S.; Lee, J.L.; Mumtaz, F.; Elbeltagi, A.; Waqas, M.S.; Dilawar, A. Analysis of Seasonal Variations in Surface Water Quality over Wet and Dry Regions. *Water* **2022**, *14*, 1058. [CrossRef]
7. Waqas, M.; Cheema, M.J.M.; Hussain, S.; Daccache, A.; Ullah, M.K.; Iqbal, M.M.; Hussain, H. Impact of advanced and delayed irrigation practices on physico-chemical attributes of potato tubers under bed sowing. *J. Anim. Plant Sci.* **2022**, *32*, 705–720.
8. Waqas, M.S.; Cheema, M.J.M.; Waqas, A.; Hussain, S. Enhancing water productivity of potato (*Solanum tuberosum* L.) through drip irrigation system. In Proceedings of the 2nd International Conference on Horticultural Science, Faisalabad, Pakistan, 18–20 February 2016; pp. 18–20.
9. Hussain, S.; Malik, S.; Cheema, M.J.M.; Ashraf, M.U.; Waqas, M.S.; Iqbal, M.M.; Ali, S.; Anjum, L.; Aslam, M.; Afzal, H. An overview on emerging water scarcity challenge in Pakistan, its consumption, causes, impacts and remedial measures. *Big Data Water Resour. Eng. (BDWRE)* **2020**, *1*, 22–31. [CrossRef]
10. Kumawat, A.; Yadav, D.; Samadharmam, K.; Rashmi, I. Soil and water conservation measures for agricultural sustainability. In *Soil Moisture Importance*; Books on Demand: London, UK, 2020; pp. 23–44.
11. Evrard, O.; Vandaele, K.; van Wesemael, B.; Bielders, C.L. A grassed waterway and earthen dams to control muddy floods from a cultivated catchment of the Belgian loess belt. *Geomorphology* **2008**, *100*, 419–428. [CrossRef]
12. Gachene, C.K.; Nyawade, S.O.; Karanja, N.N. *Soil and Water Conservation: An Overview*; Zero Hunger; Springer: Cham, Switzerland, 2020; pp. 810–823.
13. Wang, T.; Hou, J.; Li, P.; Zhao, J.; Li, Z.; Matta, E.; Ma, L.; Hinkelmann, R. Quantitative assessment of check dam system impacts on catchment flood characteristics–a case in hilly and gully area of the Loess Plateau, China. *Nat. Hazards* **2021**, *105*, 3059–3077. [CrossRef]

environmental sciences proceedings

MDPI

Proceeding Paper

Applications of Robotics and UAVs in Orchards for Fruit Picking †

Saddam Hussain [1,2,3,*], Kainat Fatima [4], Muhammad Jehanzeb Masud Cheema [3,4,5], Shoaib Rashid Saleem [4,6] and Tahir Iqbal [4,5]

1 Department of Irrigation and Drainage, University of Agriculture Faisalabad, Faisalabad 38000, Pakistan
2 Department of Biological and Agricultural Engineering, University of California Davis (UC Davis), Davis, CA 95616, USA
3 National Center of Industrial Biotechnology (NCIB) Project, PMAS-Arid Agriculture University, Rawalpindi 46000, Pakistan
4 Data-Driven Smart Decision Platform (DDSDP), PMAS-Arid Agriculture University, Rawalpindi 46000, Pakistan
5 Department of Land and Water Conservation Engineering, PMAS-Arid Agricultural University, Rawalpindi 46000, Pakistan
6 Department of Farm Machinery and Precision Engineering, PMAS-Arid Agricultural University, Rawalpindi 46000, Pakistan
* Correspondence: saddamwahla2327@gmail.com

† Presented at the 1st International Precision Agriculture Pakistan Conference 2022 (PAPC 2022)—Change the Culture of Agriculture, Rawalpindi, Pakistan, 22–24 September 2022.

Abstract: Due to the intense seasonality, high labor intensity, and high cost, picking fruit and vegetables typically requires a significant amount of personnel, material resources, and time. The fruit and vegetable picking is a critical role in the agricultural production chain. At the same time, the world is facing the challenge of an aging population. As a result, the requirements of current agricultural output cannot be addressed by using the traditional ways of picking. The robots picking have been widely utilized in the domains of fruit and vegetable production due to increases in labor productivity, picking efficiency, cost, and other aspects related to the industry. Therefore, the structural characteristics and target recognition methods for the end-effectors of picking robots will be thoroughly summarized. This study will ensure that the future direction of structural development and recognition methods that are matched with fruit and vegetable picking are more visible.

Keywords: robotics; smart farming; horticulture; UAV; fruit picking

Citation: Hussain, S.; Fatima, K.; Cheema, M.J.M.; Saleem, S.R.; Iqbal, T. Applications of Robotics and UAVs in Orchards for Fruit Picking. *Environ. Sci. Proc.* **2022**, *23*, 29. https://doi.org/10.3390/environsciproc2022023029

Academic Editor: Muhammad Umair

Published: 30 December 2022

1. Introduction

The rapid growth of the world population is the increasing challenge to achieve sustainable agriculture management as long as the human population continues to grow. It is anticipated that throughout the next few decades, the human population would increase by forty percent, reaching 9.7 billion by the year 2050. Because of this, it will be necessary to double the amount of fruit production leading to enhance the agriculture land multiple times. Despite this, it is anticipated that employment in the agricultural sector would fall by one-half by the year 2050, which will result in a shortfall of five million labor farmers. As a result, almost 10% of the world's fruit cannot be harvested, an amount that is equivalent to the yearly consumption of the European Union [1].

The harvest is a seasonal, low-paid, repetitive, and labor-intensive occupation with limited employment prospects. Older farmers are retiring, and their young ones have no interest in replacing them. Labor deficits cause harvest delays, and fruit harvested with a delay of few days degrades in quality and may lose as much as 80% of its market worth. Therefore, worldwide growers lose an estimated USD 30 billion per year in potential sales from non-harvestable fruits [2]. Consequently, crop management has evolved dramatically during the past few decades. Specifically, ground and aerial robots have been implemented in agriculture, indicating their ability to meet the rising food demand by automating previously laborious agricultural processes, such as harvesting [3]. Therefore, robotic systems have been developed to compensate the human shortage, to raise the pace of harvesting, and to enhance the efficiency.

In conventional manual harvesting, the laborers use their hands to remove leaves and branches, hold the fruit, and extract it from the plant by pulling it away, occasionally with the use of a cutting instrument. Manual harvesting requires experience; an untrained farmer may unwittingly cause damage to the plants. However, the kinematics of the human hands and body, sense of touch, and muscular power endow people with innate grabbing ability and a high degree of rapid adaptation to various crop shapes and textures for delivering the appropriate detachment force. However, human abilities are limited only by fatigue. A robotic system, on the other hand, can harvest constantly, precisely, and tirelessly with regularity. Therefore, researchers attempt to imitate human harvesting techniques, resulting in kinematic models for the movement of robotic arms and the building of sophisticated end effectors with the requisite sensors for crop manipulation [4]. In addition, recent changes in dietary needs and the production of biofuels on croplands have contributed to the existing strain on the world's food supply [5].

2. Use of Robot Picking in Agriculture

Agriculture is an issue of big data without big data. Nearly half of conventional agricultural inputs (fertilizers, insecticides, fungicides, herbicides, etc.) are often wasted because they are applied in excess or in the incorrect location (between rows rather than on plants themselves). In the future, commercial farms may be operated by robots that can detect, spray, and harvest specific pieces of fruit, even when their objectives are grapes, peppers, and apples that are the same color as the surrounding leaves as shown in Figure 1 [6]. For many crops, harvesting is the most labor-intensive task, but even proponents acknowledge that no machine has been constructed that comes close to matching human sensory motor control [7]. Robots could potentially provide a timely supply of labor in many locations where there are insufficient temporary workers available during the harvesting cycle.

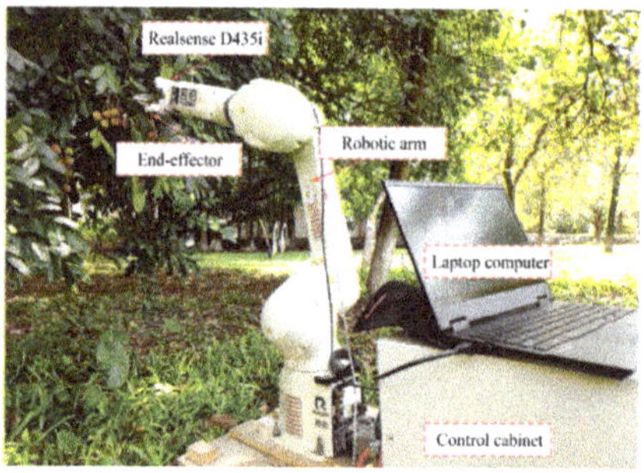

(**a**)

(**b**)

Figure 1. *Cont.*

(c) (d)

Figure 1. The use of robot in agriculture (**a**) Robot picking the Lichi [8]; (**b**) the testing of robot picking in a lab [9]; (**c**) robot picking the eggplant [10]; (**d**) proposed prototype for cotton picking [11].

3. Use of Drones for Fruit Picking

Agricultural robotic systems consist of an autonomous mobile platform, a light multi-degree-of-freedom mechanical arm, a force feedback system with a flexible end effector, a multi-sensor machine vision system, a drive control system, an intelligent decision system, and supplementary software and hardware. The arm of the drone has been programmed to grasp the desired object. A position is assigned to the drone, which is watched by the controller (human) via the camera attached to the drone. Once the location of the drone has been established according to the controller's instructions, the drone's speed is slowed, and the end effector opens and grips the desired object using suction cups attached to the effector's inner lining. This procedure is repeated until the controller obtains the desired object through a number of trials. The working environment of a fruit harvesting drone's visual components is quite complex. The object items vary in size, form, type, and surface roughness. Background and sunlight of the vegetation continuously alters. Vision-based harvesting robots must be able to sense and adapt to diverse crop varieties or environmental changes, gather information, detect targets, and train autonomously. Additionally, the robots should be capable of sophisticated reasoning and decision-making. It is a clever machine for human–computer interaction. Additionally, the robotic system should have a network transmission feature for transferring cropped photographs to a data center or server.

4. Conclusions

Today, technology is essential in every sector of society's growth, from construction to transportation to aerospace to communications to defense. Even time-honored industries such as farming need technology (in this case, smart farming) to increase output while decreasing labor requirements. However, smart farming necessitates substantial outlays of capital, enhanced coverage and connectivity, and more bandwidth to process the massive amounts of data generated by a huge number of remotely installed sensors and devices. The first steps toward a robotic future in horticulture and precision agriculture are the creation of a vision and the prediction of possible outcomes. Predicting the rate at which farmers and businesses will embrace new technologies requires a clear vision of the future, some thought about whether or not that future is desirable, and some research into historical data. Understanding what is being predicted and how these predictions offer for democratic interaction with the farmers who are meant to participate in technology transitions is becoming increasingly important. In this study, we adopt the assemblage approach to emphasize the importance of recognizing the intricate material entanglements within which anticipatory assembly occurs in order to include farmers as active actors in technology transitions.

Author Contributions: Conceptualization, S.H., K.F. and M.J.M.C.; methodology, S.H., K.F. and S.R.S.; validation, T.I.; formal analysis, S.H., and K.F.; investigation, S.R.S, M.J.M.C. and T.I.; resources, S.H. and M.J.M.C.; data curation, S.R.S., M.J.M.C. and T.I.; writing—original draft preparation, S.H. and K.F.; writing—review and editing, S.H. and K.F.; visualization, S.R.S. and T.I.; supervision, S.H., M.J.M.C. and S.R.S.; project administration, S.H., S.R.S., M.J.M.C. and T.I. All authors have read and agreed to the published version of the manuscript.

Funding: This research received no external funding.

Institutional Review Board Statement: Not Applicable.

Informed Consent Statement: Not Applicable.

Data Availability Statement: Not Applicable.

Conflicts of Interest: The authors declare no conflict of interest.

References

1. Liu, J.; Ling, C.; Guo, S.C.; Peng, Z.Y. Design and research of citrus fruit picking machine. *Packag. Eng.* **2019**, *40*, 56–62.
2. Zhuang, J.J.; Luo, S.M.; Hou, C.J.; Tang, Y.; He, Y.; Xue, X.Y. Detection of orchard citrus fruits using a monocular machine vision-based method for automatic fruit picking applications. *Comput. Electron. Agric.* **2018**, *152*, 64–73. [CrossRef]
3. Zhao, Y.; Gong, L.; Liu, C.; Huang, Y. Dual-arm robot design and testing for harvesting tomato in greenhouse. *IFAC-PapersOnLine* **2016**, *49*, 161–165. [CrossRef]
4. Berners-Lee, M.; Kennelly, C.; Watson, R.; Hewitt, C.N. Current global food production is sufficient to meet human nutritional needs in 2050 provided there is radical societal adaptation. *Elem. Sci. Anthr.* **2018**, *6*, 52. [CrossRef]
5. Morar, C.; Doroftei, I.; Doroftei, I.; Hagan, M. Robotic applications on agricultural industry. A review. In Proceedings of the 9th International Conference on Advanced Concepts in Mechanical Engineering—ACME 2020, Iași, Romania, 4–5 June 2020; Volume 997, p. 012081.
6. Godfray, H.C.J.; Beddington, J.R.; Crute, I.R.; Haddad, L.; Lawrence, D.; Muir, J.F.; Pretty, J.; Robinson, S.; Thomas, S.M.; Toulmin, C. Food security: the challenge of feeding 9 billion people. *Science* **2010**, *327*, 812–818. [CrossRef] [PubMed]
7. Arikapudi, R.; Vougioukas, S.G.J.I.A. Robotic tree-fruit harvesting with telescoping arms: A study of linear fruit reachability under geometric constraints. *IEEE Access* **2021**, *9*, 17114–17126. [CrossRef]
8. Wang, H.; Lin, Y.; Xu, X.; Chen, Z.; Wu, Z.; Tang, Y. A study on long-close distance coordination control strategy for Litchi picking. *Agronomy* **2022**, *12*, 1520. [CrossRef]
9. Chen, W.; Xu, T.; Liu, J.; Wang, M.; Zhao, D. Picking robot visual servo control based on modified fuzzy neural network sliding mode algorithms. *Electronics* **2019**, *8*, 605. [CrossRef]
10. Gonzalez-de-Santos, P.; Fernández, R.; Sepúlveda, D.; Navas, E.; Emmi, L.; Armada, M. Field robots for intelligent farms—Inhering features from industry. *Agronomy* **2020**, *10*, 1638. [CrossRef]
11. Fue, K.G.; Porter, W.M.; Barnes, E.M.; Rains, G.C. An extensive review of mobile agricultural robotics for field operations: focus on cotton harvesting. *AgriEngineering* **2020**, *2*, 150–174. [CrossRef]

environmental sciences proceedings

MDPI

Proceeding Paper

Application Predictive Control Strategies Based on Models for Optimal Irrigation of Andean Crops †

Iván Beltrán Ccama [1,*] and José Oliden Semino [2]

1 Departamento de Ingeniería Química, Facultad de Química e Ingeniería Química, Universidad Nacional Mayor de San Marcos, Lima 15081, Peru
2 Departamento de ingeniería, Universidad Tecnológica del Perú, Lima 15487, Peru
* Correspondence: ivan.beltran@unmsm.edu.pe

† Presented at the 1st International Precision Agriculture Pakistan Conference 2022 (PAPC 2022)—Change the Culture of Agriculture, Rawalpindi, Pakistan, 22–24 September 2022.

Abstract: Irrigation for high Andean agriculture is traditionally performed with rainwater and without the use of technology, where the influence of changes in water volumes and/or water losses is not considered. Likewise, the limited information on high Andean crops generates a lag in the use of irrigation technology. Improving the efficiency of irrigation in crops contributes substantially to the sustainable use of water. One way to perform this is by applying control strategies to irrigation processes that consider implementing a feedback logic of the water necessary for irrigation, thus satisfying the water demand of plants and minimizing waste. The article proposes a control strategy applying a model predictive control (MPC) that calculates the optimal amount of water for daily irrigation. The most important attraction of the model is the prediction and future behavior of the controlled variables as a function of the changes in the manipulated variables. The objective is to improve the productivity of the crop at minimum water consumption. For this, it will be necessary to use models that link with the Aquacrop software and which are allowed to be a source of data, as well as being used for the prediction of future values. The predictive model is evaluated in the Quinoa crop (*Chenopodium Quinoa Willdenow*), and the information is validated against the traditional irrigation data existing in the literature. Preliminary results indicate that the predictive model can achieve greater crop efficiency and reduce significant irrigation water supplies.

Keywords: model predictive control; precision irrigation; Quinoa; system identification

Citation: Ccama, I.B.; Semino, J.O. Application Predictive Control Strategies Based on Models for Optimal Irrigation of Andean Crops. *Environ. Sci. Proc.* **2022**, *23*, 30. https://doi.org/10.3390/environsciproc2022023030

Academic Editor: Muhammad Naveed Tahir

Published: 3 January 2023

1. Introduction

Agriculture is the sector that consumes the most water on the planet, representing approximately 70% of total use in 2020 [1]. An alternative to achieve the best use of water resources in agriculture is to apply control strategies with proven performance in the industry. Classic control methodologies, such as on/off control and proportional integral derivative (PID) control, are easy to implement and have been proven to be efficient. However, they are limited and do not consider the specific conditions of the crop, given the complexity of agricultural systems (non-linearity, multivariate). MPC strategies have shown superior performance compared to classical control strategies [2,3]. This controller is based on three ideas: the use of a prediction model, the optimization on a sliding horizon, and feedback adjustment [4]. It also allows the introduction of restrictions. The literature mentions some applications of MPC in irrigation systems [5–7]. However, there is no report of application to high Andean crops, such as Quinoa (*Chenopodium Quinoa Wildenow*), among others.

In this article, a control application of MPC was implemented in a Quinoa crop to different irrigation management strategies [8,9], as well as to simulate the yield of the crop to water and evaluate the conditions in which water is a limiting factor in production [8,10].

In this investigation, the performance of the MPC controller applied to the irrigation of the Quinoa crop was evaluated. This involves taking into account AquaCrop-OpenSource (AquaCrop-OS) as a plant model and an autoregressive structure with exogenous input (ARX) as a linear prediction model. The results obtained were compared with the irrigation methods available in the Aqua-Crop-OS gallery. All simulations were performed in MATLAB 2020a.

2. Methodology

2.1. Quinoa Crop

Quinoa is a whole grain, native to the Andes of Bolivia, Chile, and Peru. This crop is tolerant to abiotic stress, hydric stress, and requires a less amount of water for its vegetative growth [11]. It has extraordinary adaptability in agro-ecological conditions from sea level to 4000 masl., being able to withstand temperatures from –4°C to 38°C and grow with relative humidity between 40% and 70% [12]. The Quinoa in traditional farming presents critical phenological stages of susceptibility and tolerance to the need for irrigation. This is reported in accordance with the *Instituto Nacional de Innovación Agraria (INIA-Puno)*, a public institution of the Peruvian government.

2.2. Aquacrop

It was developed by the FAO in order to improve water productivity in rainfed and irrigated conditions. This simulates the yield response of arable crops to water and is particularly suitable for conditions where water is a limiting factor in crop production [8]. It has been validated for various crops, such as corn [10] and Quinoa [9]. It was developed in 2009 and its open access version AquaCrop-OS is presented in [13]. The program introduces crop information according to various characteristics: climate, type of crop, irrigation, soil, and others. The results obtained are crop growth, water balance, water content in the crop, and others. The methods provided by AquaCrop-OS are rainfed, soil moisture based, fixed interval, specified time series, and net calculation.

2.3. Model Predictive Control

The usual MPC approach is described by Equations (1)–(3).

$$\min_{u(k),\ \ldots,\ u(k+N-1)} \sum_{i=0}^{N-1} \|y(k+i-1)-r(k+i-1)\|_Q^2 + \|u(k+i)\|_R^2 \tag{1}$$

Subject to:

$$y(k+1) = f(y(k),\ u(k),\ v(k)) \tag{2}$$

$$u_{min} \leq u(k) \leq u_{max}, k = 0, \ldots, N-1 \tag{3}$$

where, y is the control variable, u is the manipulable variable, v is the measurable disturbs, r is the reference, and Q and R correspond to the weight of each term of Equation (1). The function f defines the prediction model of the controller. This problem is solved at each sampling instant.

2.4. ARX Model

The ARX model is represented in the form of a difference equation for multiple inputs as follows:

$$A(z)y(k) = \begin{bmatrix} B_1(z^{-1}) & B_2(z^{-1}) & B_2(z^{-1}) \end{bmatrix} \begin{bmatrix} u(k) \\ v_1(k) \\ v_2(k) \end{bmatrix} + e(k) \tag{4}$$

where:

$$A(z) = 1 + a_1 z^{-1} + \ldots + a_n z^{-n} \tag{5}$$

$$B_i(z) = b_{i,1} z^{-1} + \ldots + b_{i,m} z^{-m} \tag{6}$$

where $y(k)$ is the system output, $u(k)$ is the system input, v_i is the measurable disturbs, $e(k)$ is the system disturbance, d is the system delay, n is the degree of $A(z)$, m is the degree of $B_i(z)$ and i is the number of inputs. Equation (4) in regression form is:

$$\begin{aligned} y(k) = a_1 y(k-1) \quad & + \ldots + a_n y(k-n) + b_{1,1} u(k-1) + \ldots + b_{1,m} u(k-m) \\ & + b_{2,1} v_1(k-1) + \ldots + b_{2,m} v_1(k-m) + b_{3,1} v_2(k-1) + \ldots \\ & + b_{3,m} v_2(k-m) + e(k) \end{aligned} \quad (7)$$

For this work, it is considered that $y(k)$ is the water deficit, $u(k)$ is the irrigation, $v_1(k)$ is the evapotranspiration, and $v_2(k)$ is the precipitation.

3. Results

The research was carried out by simulating the conditions of the Quinoa crop in the high Andean phytogeographic domain in the region of Puno, Peru. It is located between 3812 and 5500 m above sea level. The model for response to water deficit was performed by simulation in Aquacrop-OS, a pseudorandom binary sequence (PRBS) input of irrigation (mm), and real meteorological data from the years 1964–2021 taken from Servicio Nacional de Hidrología y Meteorología (*SENAMHI*), and Quinoa data obtained from INIA-Puno were used. Normalized mean square error (NMSE) was used as evaluation criterion to estimate the parameters and validation of the model, obtaining the best performance in 2017 (NMSE = 5.7212×10^{-4}) and the worst for 1984 (NMSE = 0.0164) for the identification experiment. For validation, NMSE = 5.9894×10^{-4} and NMSE = 0.0202, respectively.

Model Predictive Control in Quinoa Crop

The control variable for this work is the water deficit (mm), and the manipulable variable is irrigation (mm). The measurable disturbances are evapotranspiration (mm) and precipitation (mm). Figure 1 shows the simulation result of the MPC controller for a simulation of a growing season using an ARX structure as a prediction model. Table 1 shows the crop yield per hectare (ton/h) and the total irrigation (mm) of the simulated methods.

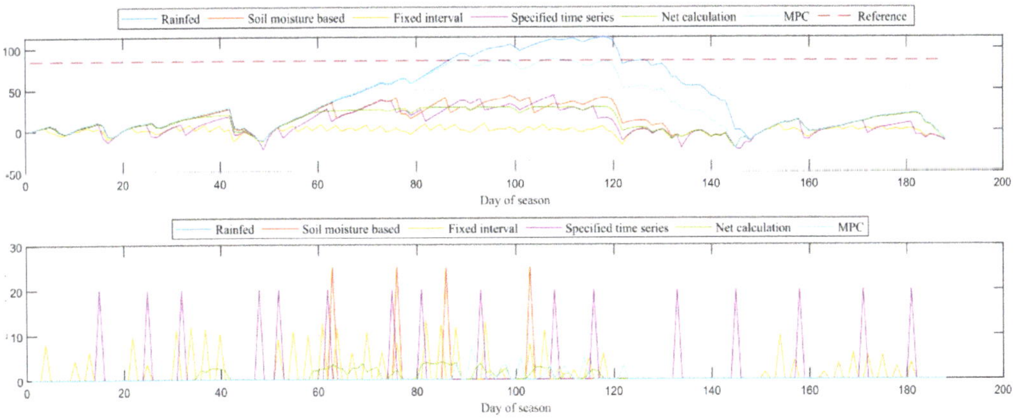

Figure 1. Comparison of the results of irrigation methods in the crop season in Puno—2017.

Table 1. Field yield and total irrigation.

Method	Field Yield (Ton/He)	Total Irrigation (mm)
Rainfed	4.31	0
Soil moisture-based	4.72	164.63
Fixed interval	4.72	289.26
Specified time series	4.72	320
Net calculation	4.72	160.08
MPC	4.72	80.17

It was observed that the method that consumes the least amount of water is the rainfed method, followed by the MPC; but, the former has the worst field yield. The MPC has similar field yield of the simulated methods with lower water consumption.

4. Conclusions

In this work, the problem of MPC control of water deficit applied in a quinoa crop model using AquaCrop-OS was presented. An ARX structure with multiple inputs and one output was proposed as a prediction model. The proposed irrigation methodology presents the best performance among the simulated methods. The future work of this research should be the implementation of the virtual model of Quinoa in a real irrigation system to determine the true efficiency and viability, according to the irrigation recommendations proposed by MPC and Aquacrop.

Supplementary Materials: The following are available online at https://www.mdpi.com/article/10.3390/environsciproc2022023030/s1, Supplementary S1.

Author Contributions: Conceptualization, I.B.C. and J.O.S.; methodology, I.B.C.; software, J.O.S.; validation, I.B.C.; formal analysis, I.B.C. and J.O.S.; investigation, I.B.C.; resources, I.B.C.; data curation, J.O.S. and I.B.C.; writing—original draft preparation, I.B.C.; writing—review and editing, I.B.C. and J.O.S.; visualization, I.B.C.; supervision, I.B.C.; project administration, I.B.C. All authors have read and agreed to the published version of the manuscript.

Funding: This research was funded by the project Concytec–Banco Mundial "Mejoramiento y Ampliación de los Servicios del Sistema Nacional de Ciencia Tecnología e Innovación Tecnológica" through its executing unit, ProCiencia. Contract 04-2018-FONDECYT/BM.

Institutional Review Board Statement: Not applicable.

Informed Consent Statement: Not applicable.

Data Availability Statement: Supplementary S1 (Supplementary Materials).

Conflicts of Interest: The authors declare no conflict of interest.

References

1. Pltonykova, H.; Koeppel, S.; Bernardini, F.; Tiefenauer-Linardon, S. *The United Nations World Water Development Report 2020: Water and Climate Change*; UNESCO: Paris, France, 2020.
2. Lopez-Jimenez, J.; Vande Wouwer, A.; Quijano, N. Dynamic Modeling of Crop–Soil Systems to Design Monitoring and Automatic Irrigation Processes: A Review with Worked Examples. *Water* **2022**, *14*, 889. [CrossRef]
3. Lozoya, C.; Mendoza, C.; Mejía, L.; Quintana, J.; Quintana, G.; Bustillos, M.; Bustillos, O.; Solís, L. Model Predictive Control for Closed-Loop Irrigation. *IFAC Proc. Vol.* **2014**, *47*, 4429–4434. [CrossRef]
4. Camacho, E.; Bordons, C. Control Predictivo: Pasado, presente y futuro", Revista Iberoamericana de. *Automática E Inf. Ind. (RIAI)* **2004**, *1*, 1–28.
5. Ding, Y.; Wang, L.; Li, Y.; Li, D. Model predictive control and its application in agriculture: A review. *Comput. Electron. Agric.* **2018**, *151*, 104–117. [CrossRef]
6. Gu, Z.; Qi, Z.; Burghate, R.; Yuan, S.; Jiao, X.; Xu, J. Irrigation Scheduling Approaches and Applications: A Review. *J. Irrig. Drain. Eng.* **2020**, *146*. [CrossRef]
7. Abioye, E.A.; Abidin, M.S.Z.; Mahmud, M.S.A.; Buyamin, S.; Ishak, M.H.I.; Rahman, M.K.A.I.; Otuoze, A.O.; Onotu, P.; Ramli, M.S.A. A review on monitoring and advanced control strategies for precision irrigation. *Comput. Electron. Agric.* **2020**, *173*, 105441. [CrossRef]

8. FAO. AquaCrop, el Modelo de Productividad del Agua de los Cultivos. 2016. Available online: https://www.fao.org/3/i7455s/i7455s.pdf (accessed on 26 August 2022).
9. Otiniano Mego, G.L. Calibración del Modelo Aquacrop para tres Variedades de Quinua. 2022, Tesis de Pregrado, Universidad Nacional Agraria la Molina. Available online: https://repositorio.lamolina.edu.pe/handle/20.500.12996/5427 (accessed on 15 August 2022).
10. Shirazi, S.Z.; Mei, X.; Liu, B.; Liu, Y. Assessment of the AquaCrop Model under different irrigation scenarios in the North China Plain. *Agric. Water Manag.* **2021**, 257. [CrossRef]
11. Apaza, V.; Cáceres, G.; Estrada, R.; Pinedo, R. Catalogue of Commercial Varieties of Quinoa in Peru. 2015. Available online: www.fao.org/publications (accessed on 15 August 2022).
12. CIRAD; FAO. State of the Art Report on Quinoa around the World in 2013–2015. Available online: http://www.fao.org/3/contents/ca682370-10f8-40c2-b084-95a8f704f44d/i4042e00.htm (accessed on 12 July 2022).
13. Foster, T.; Brozović, N.; Butler, A.P.; Neale, C.M.U.; Raes, D.; Steduto, P.; Fereres, E.; Hsiao, T.C. AquaCrop-OS: An open source version of FAO's crop water productivity model. *Agric. Water Manag.* **2017**, *181*, 18–22. [CrossRef]

environmental sciences proceedings

MDPI

Proceeding Paper

Remote Sensing in Precision Agriculture for Irrigation Management †

Tayyaba Samreen [1,*], Muhammad Ahmad [1], Muhammad Talha Baig [1], Sehrish Kanwal [2], Muhammad Zulqernain Nazir [1] and Sidra-Tul-Muntaha [3]

1 Institute of Soil and Environmental Sciences, University of Agriculture, Faisalabad 38000, Pakistan
2 Data-Driven Smart Decision Platform, PMAS-Arid Agriculture University, Rawalpindi 46000, Pakistan
3 Directorate of Pest Warning and Quality Control of Pesticides, Punjab, Lahore 54000, Pakistan
* Correspondence: tybasamreen@gmail.com or tayyaba.samreen@uaf.edu.pk
† Presented at the 1st International Precision Agriculture Pakistan Conference 2022 (PAPC 2022)—Change the Culture of Agriculture, Rawalpindi, Pakistan, 22–24 September 2022.

Abstract: The ever-increasing world's population, the consequent heavy demand for food supply, and the lack of rain-fed agriculture to meet such demands have increased the role of irrigation in agricultural production. Water management in the irrigation system is the biggest problem, particularly in regions where the effects of climate change are noticeable. The amount and timing of the crop's water requirements are under debate. Information about crops, weather, and/or soil is needed for this purpose. Unfortunately, getting such information is difficult, especially when working with enormous tracts of property. Scientists have been working to find the answers to these issues for many years. The use of remote sensing to gather the necessary data is one area that has attracted interest. The advantage of remote sensing is that data collection over vast distances becomes simple and efficient. For the purpose of enhancing sustainability, crop yield, and environmental quality, precision agriculture involves the use of agronomic concepts and innovative technology to control the geographical and temporal variance related to every aspect of agricultural output. Crop water status is monitored in agricultural areas using a variety of remote sensing techniques. It mainly includes remote sensing, crop monitoring in terms of water management, the use of drones, and modern irrigation techniques for the purpose of saving water and increasing water use efficiency in crops. This review is focused on remote sensing technologies for precision irrigation used to calculate evapotranspiration, infrared thermography, crop water status, and crop attributes. The framework for achieving this goal has included precision agriculture as a valuable element. By using these techniques, we can overcome the problem of water shortage, which is crucial for agriculture.

Keywords: precision agriculture; remote sensors; crop water status; irrigation management

Citation: Samreen, T.; Ahmad, M.; Baig, M.T.; Kanwal, S.; Nazir, M.Z.; Sidra-Tul-Muntaha Remote Sensing in Precision Agriculture for Irrigation Management. *Environ. Sci. Proc.* **2022**, 23, 31. https://doi.org/10.3390/environsciproc2022023031

Academic Editor: Shoaib Rashid Saleem

Published: 3 January 2023

1. Introduction

Precision farming is at the center of offering solutions to the industry's major issues. Precision agriculture is defined as the "application of technologies that integrate sensors, information systems, improved machinery, and versed management to optimize productivity within sustainable agricultural systems". The main goals of precision farming include economic viability, environmental protection, and sustainability. It is considered a system that assists in reorganizing the total agricultural system in the direction of sustainable, low-input, and high-efficiency agriculture [1]. However, it is still difficult to determine how much water is needed to run various irrigation management systems for a wide variety of crops. To meet the certain irrigation requirements of each crop field on the farm, irrigation systems are selected, created, and managed through the use of sensors. We must better manage our water resources if we want to produce enough food to fulfill future demands in an era of decreasing water availability. Good information is necessary for management and planning irrigation systems, yet it is currently difficult to find credible information on how to use water resources. It is not a simple task to provide accurate information at scales ranging from farmer fields complete river basins, including millions of hectares of irrigated land. However, general information about the agricultural and hydrological characteristics of the land surface may be obtained for large areas using space-borne remote sensing measures [2]. During the last 20 years, there have been significant advancements in the ability of remote sensing applications to recognize and monitor crop growth and other associated biophysical characteristics; however, there are still a number of problems that need to be fixed. Depending on the capabilities of data gathering and analysis, remote sensing techniques are now recognized as efficient and effective measurements for irrigation water management [3]. The development of technologies such as remote sensing, mobile computing, telemetry, and satellite monitoring has greatly aided the solution of the water management problem [4]. For various crops, it is crucial to maintain a powerful connection between farm water conditions and crop production [5]. The exercise of remote sensing techniques has significantly enhanced the characterization of surface water bodies, the forecasting of rainfall and temperature, the estimation of soil moisture, soil surface characteristics, and evapotranspiration. With the use of high-resolution satellite data, it is now feasible to monitor flood, drought, and irrigation management events in close to real time [5].

2. Remote Sensing for Water Management

Remote sensing systems employing information and communication technologies generally produce enormous amounts of spectrum data due to the high spatial, temporal, and radiometric resolutions necessary for applications in precision agriculture [6].

2.1. Remote Sensors

Modern agriculture is increasingly relying on the remote sensors for improved crop production with the efficient use of resources. The energy released by crops is calculated by thermal infrared sensors to determine the temperature, which is used to determine the irrigation needs and crop water shortages. Microwave sensors operate similarly to thermal sensors in order to measure the estimated energy from the ground surface. These are mostly utilized for large-scale irrigation of crops and soil moisture measurements [7]. A unique sustainable agriculture method enables the supply of water to the plant in small, regulated dosages at the proper times and locations to ensure the best possible growing circumstances [7].

2.2. Modern Irrigation Technology

A crucial component in adjusting the water needs of crops is the use of sensors to monitor crop water status. Numerous soil scientists see soil variables, such as soil water content, as an essential component of scheduling tools for regulating irrigation [8]. For checking or detecting the water content in a soil network of wireless sensors, various mobile applications and software were built. The design of wireless sensor networks was created, and the new irrigation management system was developed using this network as a basis. With the assistance of data analysts, the irrigation test is achieved by using real-time moisture data and expert data. The smart irrigation technologies include monitoring devices and different types of sensors, such as some mobile applications, temperature sensors, and soil moisture sensors, as shown in Figure 1. This technology is considered effective and practicable for use in the fields of precision agriculture and sustainable water resources.

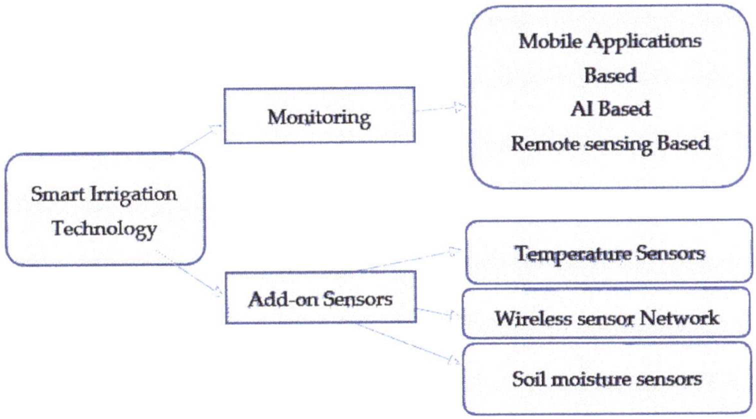

Figure 1. Smart Irrigation Technologies.

Nevertheless, water scarcity, inadequate irrigation infrastructure, low water quality, inadequate financial resources, limited farmer involvement in decision-making, and fewer extension services are the main obstacles preventing the development of modern irrigation practices [8]. To improve irrigation management, farmers must monitor data such as soil type, soil pH, soil moisture, and nutrients, to make decisions that reduce agricultural problems. In order to address the complex issues relevant to agriculture, the current methodology and developing technologies must be integrated into a data-driven technology to achieve proper irrigation [8].

2.3. Drones

More recent data may be provided by a drone, which allows for great precision in determining water usage issues that are not seen clearly on the ground level. Therefore, to upgrade crop quality, production, and profitability, precision farming builds on the use of cutting-edge technologies, including field mapping and satellite imaging. Additionally, it makes the best use of conventional resources. Huang [6] created a cutting-edge drone with an infrared camera to examine the agricultural area and illustrate the difference between healthy and infected crops [8]. Drone surveillance of greenhouses can boost agricultural production by detecting insect assaults earlier, which lowers the trip expenses over extremely vast distances and fixes any irrigation problems found on the farm. Additionally, drones and intelligent sensors might be used to provide an effective tool for agriculture in the future.

3. Conclusions

There is unquestionably a need for better management of the world's agricultural resources, such as water, due to the growing population pressure and the requirement for enhanced agricultural production. To accomplish this, it is important to gather trustworthy information on the types, quality, amount, and locations of water resources. Remote sensing technology is a crucial instrument for enhancing the current system of collecting and producing data on agriculture and natural resources. The ability to schedule irrigation and assess the effectiveness of irrigation systems using data received from remote sensing makes it a significant tool in irrigation water management. Since irrigation plays a significant role in agricultural output, the new emerging technologies of remote sensing and modern irrigation techniques will prove to be very fruitful for water management in vast irrigation systems and enhancing the water use efficiency of crops.

Author Contributions: Conceptualization, guiding and improving the article, T.S.; writing the original draft, M.A. and M.T.B.; conceptualization, drafting and revising the article S.K., M.Z.N. and S.-T.-M. All authors have read and agreed to the published version of the manuscript.

Funding: This research received no external funding.

Institutional Review Board Statement: Not applicable.

Informed Consent Statement: Not applicable.

Data Availability Statement: No datasets were generated or analyzed for this article.

Acknowledgments: The authors are thankful to organizing team of the First International Precision Agriculture Pakistan conference held on 22–24 September 2022.

Conflicts of Interest: The authors declare no conflict of interest.

References

1. Bonfante, A.; Sellami, M.H.; Abi Saab, M.T.; Albrizio, R.; Basile, A.; Fahed, S.; Bouma, J. The role of soils in the analysis of potential agricultural production: A case study in Lebanon. *Agric. Syst.* **2017**, *156*, 67–75. [CrossRef]
2. Garrido-Rubio, J.; Calera, A.; Arellano, I.; Belmonte, M.; Fraile, L.; Ortega, T.; González-Piqueras, J. Evaluation of remote sensing-based irrigation water accounting at river basin district management scale. *Remote. Sens.* **2020**, *12*, 3187. [CrossRef]
3. Kanda, E.K.; Lutta, V.O. The status and challenges of a modern irrigation system in Kenya: A systematic review. *Irrig. Drain.* **2022**, *71*, 27–38. [CrossRef]
4. Conrad, C.; Usman, M.; Morper-Busch, L.; Schönbrodt-Stitt, S. Remote sensing-based assessments of land use, soil and vegetation status, crop production and water use in irrigation systems of the Aral Sea Basin. A review. *Water Secur.* **2021**, *11*, 100078. [CrossRef]
5. Alvino, A.; Marino, S. Remote sensing for irrigation of horticultural crops. *Horticulturae* **2017**, *3*, 40. [CrossRef]
6. Huang, Y.; Chen, Z.X.; Tao, Y.U.; Huang, X.Z.; Gu, X.F. Agricultural remote sensing big data: Management and applications. *J. Integr. Agric.* **2018**, *17*, 1915–1931. [CrossRef]
7. Garrido-Rubio, J.; Gonzalez-Piqueras, J.; Campos, I.; Osann, A.; Gonzalez-Gomez, L.; Calera, A. Remote sensing–based soil water balance for irrigation water accounting at plot and water user association management scale. *Agric. Water Manag.* **2020**, *238*, 106236. [CrossRef]
8. Abdelhaleem, F.S.; Basiouny, M.; Ashour, E.; Mahmoud, A. Application of remote sensing and geographic information systems in irrigation water management under water scarcity conditions in Fayoum, Egypt. *J. Environ. Manag.* **2021**, *299*, 113683. [CrossRef] [PubMed]

environmental sciences proceedings

MDPI

Proceeding Paper

Remote Sensing for Precise Nutrient Management in Agriculture †

Tayyaba Samreen [1,*], Sidra Tahir [1], Samia Arshad [1], Sehrish Kanwal [2], Faraz Anjum [1], Muhammad Zulqernain Nazir [1] and Sidra-Tul-Muntaha [3]

1 Institute of Soil and Environmental Sciences, University of Agriculture Faisalabad, Faisalabad 38040, Pakistan
2 Data-Driven Smart Decision Platform (DDSDP), PMAS University of Arid Agriculture Rawalpindi, Rawalpindi 46000, Pakistan
3 Directorate of Pest Warning and Quality Control of Pesticides, Punjab, Lahore 54810, Pakistan
* Correspondence: tybasamreen@gmail.com or tayyaba.samreen@uaf.edu.pk

† Presented at the 1st International Precision Agriculture Pakistan Conference 2022 (PAPC 2022)—Change the Culture of Agriculture, Rawalpindi, Pakistan, 22–24 September 2022.

Abstract: Agricultural sustainability and food security are adversely affected by nutrient deficiency in the soil that in turn reduces crop yield. To restore soil fertility, precision agriculture (PA) techniques are highly encouraged. The PA techniques include the use of integrated sensors, information systems, better-quality machinery, and informed management to improve productivity. The quality and quantity of agricultural products can be improved by precision farming. The use of remote sensing is a nondestructive technique that facilitates the application of PA. The nutrient use efficiency of crops can be improved by using PA technology. In this regard, various remote sensing techniques including hyperspectral remote sensing, visible light remote sensing, and the back-propagation neural network (BPNN) model combined with ordinary kriging (OK) known as BPNKOK are currently being employed to improve soil nutrients management. These techniques assist in non-destructive monitoring of plant growth and hence aid in sustaining crop yields.

Keywords: precision agriculture; remote sensing; nutrient management; soil fertility

Citation: Samreen, T.; Tahir, S.; Arshad, S.; Kanwal, S.; Anjum, F.; Nazir, M.Z.; Sidra-Tul-Muntaha Remote Sensing for Precise Nutrient Management in Agriculture. *Environ. Sci. Proc.* **2022**, *23*, 32. https://doi.org/10.3390/environsciproc2022023032

Academic Editor: Shoaib Rashid Saleem

Published: 3 January 2023

1. Introduction

Precision agriculture (PA) is a farming system, based on integrated information and farm production, that helps us to enhance the production of farms and ensure site-specific nutrient application. The continuous use of chemical fertilizers has resulted in detrimental impacts on soil quality and the overall environment. The use of PA technology offers many advantages such as those related to the economy, environmental sustainability, and overall society [1]. Over the last 10 years, global investments and improvements in PA technology have increased dramatically [2]. PA has played a role in different fields such as sustainability, quality of several crops, productivity, protection of the environment, rural–urban agriculture development, and farm food quality [3].

The depletion of soil fertility and productivity is another main concern in the growing world. The deficiency of different nutrients including nitrogen, sulfur, potassium, zinc, and boron is the main problem in the soil. In Asian soils, nitrogen deficiency is more common. In some areas, potassium deficiency is also increasing. Sulfur is found deficient in more than 40% of samples. The deficiency of zinc in soil is up to 49%, about 12% of soils do not possess the required amount of iron and 5% of soils are deficient in copper and manganese. Moreover, 33% of boron-deficient soils were observed in 36,800 samples of soil. The efficiency of agricultural systems can be increased by using efficient agricultural production technologies such as remote sensing. By using these techniques, information on soil fertility can be collected without disturbing its surface layer.

2. Remote Sensing Applications

Remote sensing (RS), along with other systems like GPS (global positioning system) and GIS (geographic information systems), is used to access and control the agricultural activities. RS techniques are used in agriculture to determine crop and field status such as, soil moisture contents, soil fertility status, detection of crop stress and disease-causing pests for sustainable production of crops and to improve the economic status of the country [4]. The hyperspectral images, obtained from the solar-induced chlorophyll fluorescence (SIF), are used to estimate the plant nutrient and photosynthetic status [5]. Another sensor, such as vegetation indices (VI), is used for crop nutritional status valuation [6].

2.1. Role of Remote Sensing in Nutrient Management

Remote sensing is the science that involves the collection and interpretation of information or data about an object by the use of remote sensors, without making any physical contact with that objects i.e., at some distance from the earth's surface [7]. RS has importance in the classification and evaluation of crops and yield. Remote sensing is the science and art of collecting data from the earth's surface without contacting it directly [8]. It is a non-destructive technique of collecting data on the earth's surface. The information can be obtained analytically over a huge geographical area, except for the observation of a single point. RS collects data from areas that are unapproachable to humans. It is autonomous from the information collected elsewhere, in contrast to other mapping sciences like GIS and cartography. The GIS and RS techniques help in improving the nutritional status of significant agricultural areas. The use of GIS and RS can reduce the costs for farmers and enhance the efficiency of fertilizers for crops that ultimately reduce nutritional stress by using site-specific nutrient management techniques [9].

The under- and over-fertilization of soil, prominent due to the heterogeneous nature of soil, can be improved by using the standard single-rate N fertilization technique. The use of variable-rate nitrogen fertilizer (VRF) can improve the effectiveness of N fertilization [10]. This VRF technique can improve nutrient use efficiency and crop output and decrease the nutrient loss from the fields, thus controlling environmental pollution [11].

2.2. Types of Remote Sensing

Different types of remote sensing techniques are used for the fast diagnosis of crop nutrition and growth. Canopy color analysis, hyperspectral remote sensing, and visible light remote sensing are useful techniques that are widely used. These are likely to become hypothetical and non-destructive diagnostic methods for nitrogen crop nutrition in the new era due to the benefits of stable, fast, accessible, and non-destructive results, as well as the good correlation between the color parameter of the NRI canopy and N content of crop and performance index. It is claimed for non-destructive analysis of N availability and its viability to evaluate the N content and plant growth readily and in real-time [12]. To measure total soil nitrogen, soil obtainable P and K content, hyperspectral images were used with a neural network model in a back-propagation neural network (BPNN) and combined ordinary kriging (OK) (Figure 1). The application of hyperspectral imaging with the BPNNOK model has proven to be an effective technique for the detection of soil nutrients [13].

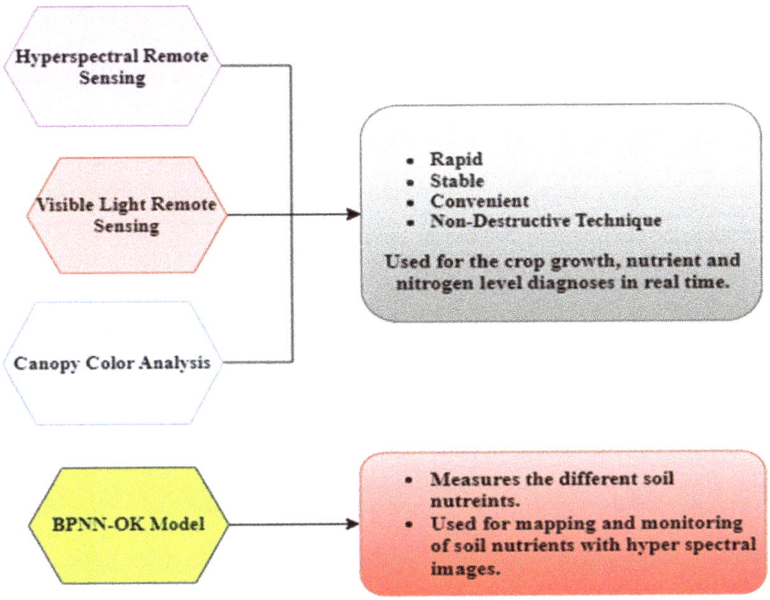

Figure 1. Types of remote sensing in nutrient management.

3. Conclusions

With the increase in the world's population and the pressure on food, there is a need to improve nutrient management through several nondestructive techniques. There are several remote sensing techniques available, such as canopy color analysis, hyperspectral remote sensing, and visible light remote sensing. In addition, the application of the BPNNOK model has been proven to be an effective technique for the detection of soil nutrients. Thus, it is concluded that the application of remote sensing is a precise technique in nutrient management that ultimately aids in improving plant growth and nutrient status.

Author Contributions: Conceptualization, guiding and improving the article, T.S.; Writing the original draft, S.T., S.A. and F.A.; conceptualization, drafting and revising the article, S.K., M.Z.N. and S.-T.-M. All authors have read and agreed to the published version of the manuscript.

Funding: This research received no external funding.

Institutional Review Board Statement: Not applicable.

Informed Consent Statement: Not applicable.

Data Availability Statement: No datasets were generated or analyzed for this article.

Acknowledgments: The authors are thankful to 1st international precision agriculture Pakistan conference held on 22–24 September 2022.

Conflicts of Interest: The authors declare no conflict of interest.

References

1. Auernhammer, H. Precision farming the environmental challenge. *Comput. Electron. Agric.* **2001**, *30*, 31–43. [CrossRef]
2. Schellberg, J.; Hill, M.J.; Gerhards, R.; Rothmund, M.; Braun, M. Precision agriculture on grassland: Applications, perspectives and constraints. *Eur. J. Agron.* **2008**, *29*, 59–71. [CrossRef]
3. Liaghat, S.; Balasundram, S.K. A review: The role of remote sensing in precision agriculture. *Am. J. Agric. Biol. Sci.* **2010**, *5*, 50–55. [CrossRef]

4. Singh, P.; Pandey, P.C.; Petropoulos, G.P.; Pavlides, A.; Srivastava, P.K.; Koutsias, N.; Bao, Y. Hyperspectral remote sensing in precision agriculture: Present status, challenges, and future trends. In *Hyperspectral Remote Sensing: Theory and Applications*; Pandey, P.C., Srivastava, P.K., Heiko Balzter, H., Bhattacharya, B., Eds.; Elsevier: Amsterdam, Netherlands, 2020; pp. 121–146. ISBN 9780081028940.
5. Steven, M.D.; Clark, J.A. Applications of remote sensing in agriculture-A Review. *Int. J. Current Microbiol. App. Sci.* **2013**, *8*, 2270–2283.
6. Chang, C.Y.; Zhou, R.; Kira, O.; Marri, S.; Skovira, J.; Gu, L.; Sun, Y. An Unmanned Aerial System (UAS) for concurrent measurements of solar-induced chlorophyll fluorescence and hyperspectral reflectance toward improving crop monitoring. *Agric. For. Meteorol.* **2020**, *294*, 108145. [CrossRef]
7. Khanal, S.; Fulton, J.; Shearer, S. An overview of current and potential applications of thermal remote sensing in precision agriculture. *Comput. Electron. Agric.* **2017**, *139*, 22–32. [CrossRef]
8. Jensen, J.R. *Remote Sensing of the Environment: An Earth Resource Perspective*, 2nd ed.; Prentice Hall: Upper Saddle River, NJ, USA, 2007; pp. 1–28.
9. Justice, C.O.; Townshend, J.R.G.; Vermote, E.F.; Masuoka, E.; Wolfe, R.E.; Saleous, N.; Morisette, J.T. An overview of MODIS Land data processing and product status. *Remote. Sens. Environ.* **2002**, *83*, 3–15. [CrossRef]
10. Casa, R.; Cavalieri, A.; Cascio, B.L. Nitrogen fertilisation management in precision agriculture: A preliminary application example on maize. *Ital. J. Agron.* **2011**, *6*, e5. [CrossRef]
11. Hedley, C. The role of precision agriculture for improved nutrient management on farms. *J. Sci. Food Agric.* **2015**, *95*, 12–19. [CrossRef] [PubMed]
12. Feng, D.; Xu, W.; He, Z.; Zhao, W.; Yang, M. Advances in plant nutrition diagnosis based on remote sensing and computer application. *Neural Comput. Appl.* **2020**, *32*, 16833–16842. [CrossRef]
13. Song, Y.Q.; Zhao, X.; Su, H.Y.; Li, B.; Hu, Y.M.; Cui, X.S. Predicting spatial variations in soil nutrients with hyperspectral remote sensing at regional scale. *Sensors* **2018**, *18*, 3086. [CrossRef] [PubMed]

environmental sciences proceedings

MDPI

Proceeding Paper

Digital Twin Greenhouse Technologies for Commercial Farmers †

Kainat Fatima [1,*], Saddam Hussain [2,3,4], Shahid Ali Khan [5], Muhammad Azam Khan [1,6], Shoaib Rashid Saleem [1,7] and Sehrish Kanwal [1]

1 Data-Driven Smart Decision Platform (DDSDP), PMAS-Arid Agriculture University, Rawalpindi 46000, Pakistan
2 Department of Irrigation and Drainage, University of Agriculture Faisalabad, Faisalabad 38000, Pakistan
3 Department of Biological and Agricultural Engineering, University of California Davis (UC Davis), Davis, CA 95616, USA
4 National Center of Industrial Biotechnology (NCIB) Project, PMAS-Arid Agriculture University, Rawalpindi 46000, Pakistan
5 Directorate of Planning & Development, PMAS-University of Arid Agriculture, Rawalpindi 46000, Pakistan
6 Department of Horticulture, Faculty of Crop and Food Sciences, PMAS-University of Arid Agriculture, Rawalpindi 46000, Pakistan
7 Department of Farm Machinery and Precision Engineering, Faculty of Agricultural Engineering and Technology, PMAS-Arid Agricultural University, Rawalpindi 46000, Pakistan
* Correspondence: kainatfatima@uaar.edu.pk
† Presented at the 1st International Precision Agriculture Pakistan Conference 2022 (PAPC 2022)—Change the Culture of Agriculture, Rawalpindi, Pakistan, 22–24 September 2022.

Abstract: Technology integration between the farm and the consumer is at the heart of digital agriculture. New technologies for developing countries, such as vertical indoor farming that uses automation and robots, might speed up the elimination of rural poverty and hunger worldwide. Significant technical developments include state-of-the-art greenhouse techniques, laser-guided precision farming, AI, and blockchain. Connected farm machinery collects data that may then be used to research the soil and climate of a given area, allowing experts to offer advice on seed selection and the optimal timing for applying pesticides and fertilizers. One of the most widely used innovations of the previous century was the mobile phone. The use of digital technology will improve communication between consumers and farmers. The public now has more accessible access to information on farming because of the wealth of data collected on crops and livestock. Smart farming's impact on crop yields will be seen in the long term.

Keywords: digital twin; greenhouse horticulture; smart farming

Citation: Fatima, K.; Hussain, S.; Khan, S.A.; Khan, M.A.; Saleem, S.R.; Kanwal, S. Digital Twin Greenhouse Technologies for Commercial Farmers. *Environ. Sci. Proc.* **2022**, *23*, 33. https://doi.org/10.3390/environsciproc2022023033

Academic Editor: Tahir Iqbal

Published: 3 January 2023

1. Introduction

As a result of concerns about food security, safety, sustainability, and health, the production of horticultural goods is becoming increasingly industrialized. Greenhouses are rapidly transforming into high-tech factories that utilize a great deal of machinery and a vast variety of goods in extremely large quantities [1]. This transformation is a result of the increased use of advanced sensors and control systems for climate management, irrigation, fertigation, lighting, crop monitoring, disease scouting, harvesting, internal transportation, sorting, and packaging. On the other hand, greenhouse horticulture is becoming increasingly cutting-edge and data-driven. This phenomenon has accelerated due to technological developments such as cloud computing, the Internet of Things (IoT), big data, machine learning, augmented reality, and robots [2]. Growers now can remotely monitor and control innovative and data-driven greenhouse horticulture activities by utilizing digital information in (near) real-time. They receive notifications for potential concerns and can view a rich visual image of the plants or equipment in the greenhouse from their desk or smartphone. Growers can simultaneously mimic corrective and preventive activities on the digital depiction. Finally, the farmer can remotely implement the recommended treatments and use digital representation to confirm that the problem is solved. Without the grower's manual intervention, this intelligent management cycle will become increasingly autonomous. Finally, plants, containers, greenhouse sections, and equipment can be virtualized and remotely managed through digital twin development. It simulates the behavior of real-life objects in a virtual world [3]. The digital twin concept is new and could advance intelligent greenhouse horticulture. However, it is uncertain how much greenhouse horticulture uses digital twins. There is little expertise in creating and using digital twin-based systems for greenhouse horticulture.

2. The Present Need for Digital Agricultural Practices

The agricultural technology industry is evolving in line with other markets and becoming more knowledge-intensive. Production methods have evolved from their old forms into more efficient and creative models as a result of this shift. The farmers have been going through it recently. All of these changes may be achieved via the use of digital agriculture [4]. The concepts of "precision agriculture," which centers on agricultural production processes, and digital agriculture, which [5] characterize as an application of the digital world idea established in the 1990s, are crucial to the development of this field. When we talk about digital agriculture, we are referring to the practice of using digital and communication technology to boost agricultural productivity and longevity. The growing use of innovative, interconnected, and data-intensive computing technologies is collectively referred to as the Industry 4.0 revolution, which has brought digital agriculture and a slew of new possibilities to the agricultural sector [6]. The Global Institute of Food Security (2015) states that only approximately 20% of the world's agricultural areas are being managed using digital agriculture technology.

3. Digital Twin Concept

The idea of a digital twin has its roots in product management. There was a demand in this field for a centralized repository of product data that anybody could access at any point in the product lifecycle. As such, it was planned for the digital representation of the product to include all the necessary data for the planning, production, and upkeep of the product [7].

The IoT is predicated on the idea of interaction between digital and physical items. Every real-world item has a detailed digital counterpart that can be accessed from anywhere in the world that details the item's history, provenance, ownership, and sensory context. As IoT-based systems evolve, these digital artifacts often serve as the foundation for smart systems with highly sophisticated control features like monitoring and prediction as mentioned in Table 1. Among other reasons, the novelty of the notion leads to such intelligent systems not being presented as digital twins.

Table 1. Essential traits of digital twin technology.

Qualities	Findings	Reference
Timeliness	A digital twin represents its physical twin in (near) real-time, identifying and synchronizing state changes of the physical object and vice versa.	[8]
Fidelity	Digital twins must be unquestionably reliable and secure to be trusted for decision-making.	[8]
Integration	A digital twin integrates data from all aspects of the physical object in a unified format.	[2]
Intelligence	Digital twins can replicate products, resources, components, and processes. The digital twin can also show multiple interdependent items.	[2]

4. Digital Twins in Greenhouse Horticulture

Horticulture is characterized by a great variety and variability in production. It involves living, perishable products and production dependent on natural conditions such as weather, diseases, soil condition, seasons, and climate. This makes horticulture one of the world's most dynamic and exciting industries. To mitigate these risks, many farmers are turning to greenhouses for year-round indoor production, which provides a better-managed production environment in which temperature, fertigation, light, and moisture can all be optimized. By enabling farmers to take quick action in the face of (anticipated) deviations and to simulate treatments based on real data, digital twins may considerably boost the required control skills. Furthermore, greenhouse horticulture has become more widespread in recent years, as shown in Figure 1. Large-scale manufacturing makes the manual tracking of the growing process impossible. The growing shortage of "green labor," or skilled workers in the field of horticulture, only adds fuel to the fire.

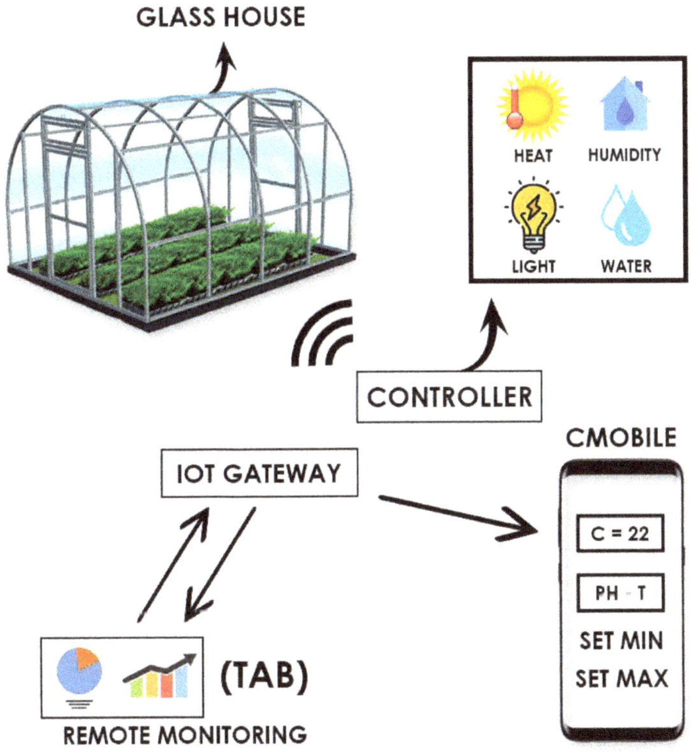

Figure 1. Design and Implementation of Integrated Control System—Sensors for "Smart" Greenhouses based on the Internet of Things (IoT) Technologies (Picture Credit: Kainat Fatima).

Digital twins may help to solve these problems by removing location, time, and human observation restrictions [9]. Remote and automated greenhouse horticulture would allow stakeholders to remotely execute, monitor, control, and coordinate greenhouse activities. This separates physical and information processes in horticulture. Sensor and energy data and data from other information owners may augment digital twins (e.g., weather data), as shown in Figure 1. Digital twins in greenhouse horticulture may also analyze past states and anticipate future behavior in terms of crop growth and yields. Thus, correctly linked digital twin apps may help farmers and stakeholders make decisions, respond instantly to predicted deviations, and remotely regulate greenhouse operations. Intelligence lets digital twins collect the implicit "green" knowledge of experienced horticulturists and learn from data. This will improve production, yields, and quality at the right moment while reducing the demand for experienced workers.

5. Conclusions

Many intricate steps work together to make agriculture. Everything needs to be broken down into manageable steps to increase productivity. How well a product does in the market depends on the farmer, technology, service and consultancy idea. As with the rest of the economy, agriculture will be digitized. The government should allocate time and money to spreading the word about the advantages of digitization among the general public. The rapid growth of e-agriculture is hindered by inadequate connectivity in rural regions, exorbitant service rates, and a lack of fundamental computer literacy and comprehension. Substantial funding is essential for developing physical infrastructure, electrical networks, broadband internet, and transportation.

Author Contributions: Conceptualization, K.F., S.H., S.A.K. and S.R.S.; methodology, K.F., S.H., and S.R.S.; validation, S.R.S., M.A.K. and S.K.; formal analysis, S.H. and K.F.; investigation, S.R.S.; resources, S.H. and S.R.S.; data curation, S.R.S.; writing—original draft preparation, S.H., K.F. and S.H.; writing—review and editing, S.H. and K.F.; visualization, S.R.S. and S. K.; supervision, S.H. and S.R.S.; project administration, S.H., S.R.S. and S.K.; All authors have read and agreed to the published version of the manuscript.

Funding: This research received no external funding.

Institutional Review Board Statement: Not applicable.

Informed Consent Statement: Not applicable.

Data Availability Statement: Not applicable.

Conflicts of Interest: The authors declare no conflict of interest.

References

1. Verdouw, C.; Tekinerdogan, B.; Beulens, A.; Wolfert, S. Digital Twins in Smart Farming. *Agric. Syst.* **2021**, *189*, 103046. [CrossRef]
2. Grieves, M.; Vickers, J. Digital Twin: Mitigating Unpredictable, Undesirable Emergent Behavior in Complex Systems. In *Transdisciplinary Perspectives on Complex Systems*; Kahlen, F., Flumerfelt, S., Alves, A., Eds.; Springer: Cham, Switzerland, 2018; pp. 85–113.
3. Sundmaeker, H.; Guillemin, P.; Friess, P.; Woelfflé, S. *Vision and Challenges for Realising the Internet of Things*; EU Publications: Luxembourg, 2010.
4. Zude-Sasse, M.; Fountas, S.; Gemtos, T.; Abu-Khalaf, N. Applications of precision agriculture in horticultural crops. *Eur. J. Hortic. Sci.* **2016**, *81*, 78–90. [CrossRef]
5. Shen, S.; Basist, A.; Howard, A. Structure of a Digital Agriculture System and Agricultural Risks Due to Climate Changes. *Agric. Agric. Sci. Procedia* **2010**, *1*, 42–51.
6. Deichmann, U.; Goyal, A.; Mishra, D. Will Digital Technologies Transform Agriculture in Developing Countries? *Agric. Econ.* **2016**, *47*, 21–33. [CrossRef]
7. Pylianidis, C.; Osinga, S.; Athanasiadis, I.N. Introducing Digital Twins to Agriculture. *Comput. Electron. Agric.* **2021**, *184*, 105942. [CrossRef]
8. Yane, D. Research and Analysis about System of Digital Agriculture Based on a Network Platform. In Proceedings of the International Conference on Computer and Computing Technologies in Agriculture, Nanchang, China, 22–25 October 2010; pp. 274–282.
9. Zhai, Z.; Martínez, J.F.; Beltran, V.; Martínez, N.L. Decision support systems for agriculture 4.0: Survey and challenges. *Comput. Electron. Agric.* **2020**, *170*, 105256. [CrossRef]

environmental sciences proceedings

MDPI

Proceeding Paper

Reshaping the Agriculture Sector of Pakistan through Innovative Agri-Tech Devices to Achieve Food Security †

Zainab Ahmed *, Ayesha Alam Khurram and Shujaat Khanzada

Department of Agriculture, Crop2x Private Ltd., Karachi 75270, Sindh, Pakistan
* Correspondence: zainab@crop2x.com

† Presented at the 1st International Precision Agriculture Pakistan Conference 2022 (PAPC 2022)—Change the Culture of Agriculture, Rawalpindi, Pakistan, 22–24 September 2022.

Abstract: Precision agriculture (PA) has the potential to radically transform agronomic systems. It is an effective approach for viable zone management in the agriculture field. In today's era of finite resources and drastic consequences of climate change, an approach to PA which is an integration of below-the-ground sensors, multispectral satellite imagery, and weather monitoring system is reshaping agriculture from static to smart. In this paper, a real-time case study at a lemon orchard in Gadap, Sindh, Pakistan is presented where PA practices are being implemented successfully. At the farm locally developed innovative agri-tech devices are deployed which are embedded with electrical conductivity, soil temperature, soil moisture sensor, and nitrogen, phosphorus, and potassium sensor to monitor real-time conditions of the soil for precision irrigation and fertilizer application. Along with device data, incorporation of weather data, agronomist advisory and use of satellite imagery offer a full-functioning monitoring system for viable decisions. This system also favors tracking variations in crop health & pest attack for precise pesticide spray. The data output is observed through a web application. Using these drivers for PA there was increased flowering in the orchard as compared to other farms in the vicinity. Hence, a promising surplus yield and least toxic better fruit quality are being obtained, along with the preservation of biodiversity and environment sustainability the output yield of lemons was quite better than the conventional agriculture practices. PA is an extraordinary approach to leap closer to food security.

Keywords: precision agriculture (PA); agri-tech devices; remote sensing; food security; Internet of Things (IoT)

Citation: Ahmed, Z.; Khurram, A.A.; Khanzada, S. Reshaping the Agriculture Sector of Pakistan through Innovative Agri-Tech Devices to Achieve Food Security. *Environ. Sci. Proc.* **2022**, *23*, 34. https://doi.org/10.3390/environsciproc2022023034

Academic Editor: Shoaib Rashid Saleem

Published: 4 January 2023

1. Introduction

Global food security drives the recent resurgence of interest in agronomic systems [1]. PA focuses on optimizing farm inputs and contributes to gap fulfillment between crop potential and productivity [2]. Together the IoT-based modules along with sensor probes, multispectral satellite imagery, and weather monitoring systems [3] constitute a management framework to achieve global food security along with the preservation of biodiversity and environment sustainability.

PA was being applied at a lemon farm in Gadap town by Crop2x an agri-tech company that locally develops IoT devices integrated with web applications to access real-time data of the field. The fertilizer for soil nutrients and pesticide treatment was considered through spatial variability with the aim to use resources efficiently and achieve high-quality fruits.

Results indicated that PA implemented in one cycle of lemon enhanced the number of flowers leading to a greater number of fruit formation, Hence farmers can improve the production of the crops by adopting PA practices to minimize soil degradation [4], conserve biodiversity, and maintain a sustainable environment to create the base of food security.

2. Method and Observation

The technological approach of PA implemented at the lemon farm aimed to reshape the agriculture practices in a structured path.

The study provided evidence that usage of resources can be reduced by accessing real-time data of the field conditions through IoT devices on web application coupled with agronomist advisory to take data-driven decisions for variable application of fertilizers, precise irrigation needed by crop for optimal growth, locating spatially low vegetation areas in field and pesticide spray management at ETL level considering weather data to maintain biodiversity and production of least toxic yield.

The highest output with better quality was achieved through PA integrating satellite imagery, fertilizer application, pesticide application, and precision irrigation/ The method is evidence that how the application of PA technology is a promising prospect for global food security and reshaping agriculture.

3. Result

3.1. Satellite Imagery

Satellite imagery by using artificial intelligence has become useful data for decision-making, in precision agriculture. The interaction of radiation with leave or canopy changes shows variability in the field through satellite imagery [5].

The low vegetation areas monitored in the lemon farm were due to the cause of pest incidents treating the localized areas reduced pesticide usage thus protecting the natural environment.

3.2. Fertilizer Application

Unknown nutrient levels in the field and refraining soil analysis leads to soil infertility. The soil sensors enable monitoring of NPK trends and other parameters [6] e.g., soil salinity, Ec, TDS and soil temperature [7].

A known quantity of NPK in soil optimized the usage and avoidance of excess fertilizer as illustrated in Figure 1. Ec functioned as a base in the selection of less alkaline fertilizers and proves to be effective for lemon trees in terms of remarkable production and maintaining soil fertility.

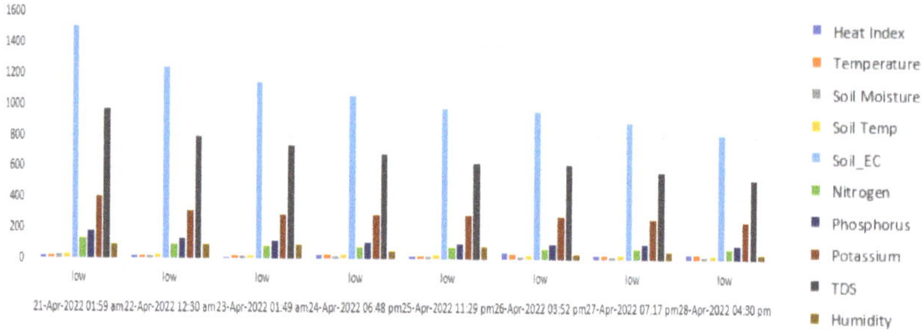

Figure 1. Real-time data of soil condition in the lemon orchard and fertilizer trend that helps in optimum inputs.

3.3. Pesticide Application

Pest surveillance in the field is influenced by favorable temperature and humidity. Integrating spatial variability and weather conditions in PA cost of pesticide application was reduced and profit was maximized.

Site-specific applications [6] of selective and bio pesticides were applied that not only contributed to pest control but also had a significant impact on natural biodiversity proven by the presence of 11 bee hives in the lemon farm. Bees responsible for pollination played a magnificent role in better flowering leading to the highest output in the vicinity.

3.4. Precision Irrigation

IoT devices equipped with soil moisture sensors were placed at certain depths for accurate measurement of the water level in the soil. A precision irrigation time was designed depending on the crop stage, root zone, geographical location, and evapotranspiration [8–10].

The system is capable to forecast irrigation required on the basis of water availability and evapotranspiration illustrated in Figure 1. This helps in eliminating excessive irrigation and root-related diseases in the orchard and irrigation management [11] contributed to good flowering and high-quality fruit production.

3.5. Soil Temperature

Plant roots growing in the soil need a conducive environment to transport nutrients and water to the plant. To keep this mechanism optimal agri-tech devices equipped with soil temperature sensors remarkably influenced soil temperature management.

At high soil temperatures organic matter leaches down the soil and the availability of nutrients through the plant roots is disturbed as the beneficial bacteria die and are unable to decompose the nutrients. At lemon, orchard mulching was done at high soil temperatures illustrated in Figure 1 to prevent the associated problems hence achieving the desired output.

4. Conclusions

Precision agriculture using agri-tech devices was implemented and practiced for a complete cycle of fruiting at a lemon farm.

In reference to the pilot project at lemon farm Gadap if the same PA technology is implemented on the other farms of the country it can provide accurate data to establish agronomic decisions. The output in precision agriculture was promising in terms of profit and the cost of inputs using remote sensing and agri-tech devices was minimized.

This technology is a breakthrough in uplifting the agricultural production of Pakistan in achieving food security.

5. Future Perspective of This Study

If, PA is practically implemented by optimizing the inputs such as water, fertilizer, and pesticide, it will result in the high production with increased quantity and quality of fruit.

Author Contributions: Conceptualization, Z.A., S.K. and A.A.K.; methodology, S.K. and A.A.K.; software, A.A.K.; validation, S.K., A.A.K. and Z.A.; formal analysis, A.A.K. and Z.A.; investigation, S.K. and A.A.K.; data curation, Z.A. and S.K.; writing—original draft preparation, S.K. and A.A.K.; writing—review and editing, Z.A., S.K. and A.A.K.; supervision, A.A.K. All authors have read and agreed to the published version of the manuscript.

Funding: This research received no external funding.

Institutional Review Board Statement: Not applicable.

Informed Consent Statement: Not applicable.

Data Availability Statement: Not applicable.

Acknowledgments: The authors are thankful to Humaira Rana for her valuable feedback.

Conflicts of Interest: The authors declare no conflict of interest.

References

1. Diaz-Ambrona, C.G.H.; Maletta, E. Achieving Global Food Security through Sustainable Development of Agriculture and Food Systems with Regard to Nutrients, Soil, Land, and Waste Management. *Curr. Sustain. Renew. Energy Rep.* **2014**, *1*, 57–65. [CrossRef]
2. Aslam, M. Agricultural productivity current scenario, constraints and future prospects in Pakistan. *Sarhad J. Agric.* **2016**, *32*, 289–303. [CrossRef]
3. Shafi, U.; Mumtaz, R.; García-Nieto, J.; Hassan, S.A.; Zaidi, S.A.R.; Iqbal, N. Precision Agriculture Techniques and Practices: From Considerations to Applications. *Sensors* **2019**, *19*, 3796. [CrossRef] [PubMed]

4. Khoso, S.; Wagan, H.; Tunio, H.; Ansari, A. An overview on emerging water scarcity in Pakistan, its causes, impacts and remedial measures. *J. Appl. Eng. Sci.* **2015**, *13*, 35–44. [CrossRef]
5. Silva, G.; Tomlinson, J.; Onkokesung, N.; Sommer, S.; Mrisho, L.; Legg, J.; Adams, I.P.; Gutierrez-Vazquez, Y.; Howard, T.P.; Laverick, A.; et al. Plant pest surveillance: From satellites to molecules. *Emerg. Top Life Sci.* **2021**, *5*, 275–287. [PubMed]
6. Rajmis, S.; Karpinski, I.; Pohl, J.P.; Herrmann, M.; Kehlenbeck, H. Economic potential of site-specific pesticide application scenarios with direct injection and automatic application assistant in northern Germany. *Precis. Agric.* **2022**, *23*, 2063–2088. [CrossRef]
7. Jacquet, F.; Jeuffroy, M.H.; Jouan, J.; Le Cadre, E.; Litrico, I.; Malausa, T.; Reboud, X.; Huyghe, C. Pesticide-free agriculture as a new paradigm for research. *Agron. Sustain. Dev.* **2022**, *42*, 8. [CrossRef]
8. Madhumathi, R.; Arumuganathan, T.; Shruthi, R. Soil NPK and Moisture analysis using Wireless Sensor Networks. Proceedings of 2020 11th International Conference on Computing, Communication and Networking Technologies (ICCCNT), Kharagpur, India, 1–3 July 2020; pp. 1–6.
9. Işık, M.F.; Sönmez, Y.; Yılmaz, C.; Özdemir, V.; Yılmaz, E.N. Precision Irrigation System (PIS) Using Sensor Network Technology Integrated with IOS/Android Application. *Appl. Sci.* **2017**, *7*, 891. [CrossRef]
10. Latif, M.; Sarwar, S. Proposal for equitable water allocation for rotational irrigation in Pakistan. *Irrig. Drain. Syst.* **1994**, *8*, 35–48. [CrossRef]
11. Ebrahim, Z.T. Is Pakistan Running Dry. In *Water Issues in Himalayan South Asia*; Palgrave Macmillan: Singapore, 2019; pp. 153–181.

environmental sciences proceedings

MDPI

Proceeding Paper

The Importance of Variable Rate Irrigation in Lowering Greenhouse Gas Emissions in the Agriculture Sector: A Review †

Saddam Hussain [1,2,3,*], Muhammad Jehanzeb Masud Cheema [3,4], Muhammad Sohail Waqas [5,6], Shoaib Rashid Saleem [4,7], Rameela Rustam [1], Muhammad Saadullah Khan [3] and Muhammad Habib Ullah [3]

1 Department of Irrigation and Drainage, University of Agriculture Faisalabad, Faisalabad 38000, Pakistan
2 Department of Biological and Agricultural Engineering, University of California Davis (UC Davis), Davis, CA 95616, USA
3 National Center of Industrial Biotechnology (NCIB) Project, PMAS-Arid Agriculture University, Rawalpindi 46000, Pakistan
4 Faculty of Agricultural Engineering and Technology, PMAS-Arid Agriculture University, Rawalpindi 46000, Pakistan
5 Soil Conservation Group, Agriculture Department (Field Wing), Government of the Punjab, Rawalpindi 46000, Pakistan
6 Department of Agricultural and Biological Engineering, Tropical Research and Education Center, IFAS, University of Florida, Homestead, FL 33031, USA
7 Center for Precision Agriculture, PMAS-Arid Agriculture University, Rawalpindi 46000, Pakistan
* Correspondence: saddamwahla2327@gmail.com; Tel.: +92-324-7712366
† Presented at the 1st International Precision Agriculture Pakistan Conference 2022 (PAPC 2022)—Change the Culture of Agriculture, Rawalpindi, Pakistan, 22–24 September 2022.

Abstract: Agriculture is extremely vulnerable to climate change, creating more difficult challenges. Presently, the agricultural sector contributes to between 19 and 29% of all global greenhouse Gas (GHG) emissions. Methane (CH_4), nitrous oxide (N_2O), and carbon dioxide (CO_2) are the main types of greenhouse gases generated by the agricultural industry. Energy use before and after farms, as well as shifting ground carbon stocks above and below as a result of changes in land use, are major sources of CO_2 emissions. There has been a trend in recent years toward lowering GHG emissions in the agriculture sector. Precision agriculture Technologies (PAT) address the field's temporal and spatial variability to maximize the usage of agricultural inputs (i.e., irrigation, fuel, and fertilizers). The PAT can keep or increase productivity while lowering GHG emissions from agricultural activities, whereas the variable rate irrigation (VRI) approach is helpful in this scenario. Recent research shows that VRI has a significant potential to mitigate GHG. The present study reviews research related to VRI that address the reduction in GHG emissions.

Keywords: agriculture; precision agriculture technologies; variable rate irrigation; drip irrigation; water management

Citation: Hussain, S.; Cheema, M.J.M.; Waqas, M.S.; Saleem, S.R.; Rustam, R.; Khan, M.S.; Ullah, M.H. The Importance of Variable Rate Irrigation in Lowering Greenhouse Gas Emissions in the Agriculture Sector: A Review. *Environ. Sci. Proc.* **2022**, *23*, 35. https://doi.org/10.3390/environsciproc2022023035

Academic Editor: Muhammad Naveed Anjum

Published: 6 January 2023

1. Introduction

Droughts, floods, and locust outbreaks are growing more common, and crop yields are decreasing; however, these abrupt climate changes are less concerning than the already existing issues for farmers worldwide [1]. Food demand is increasing as a result of variable diets, and the population is expanding in many regions of the world. The world will need to produce nearly 70% more human-required food to meet the needs of an estimated 9 billion people by 2050, making the problem of food security even more difficult [2]. The use of chemical fertilizers with pesticides and animal wastes in agricultural activities accounts for about 30% of all GHG emissions. This rate will undoubtedly continue to increase because of the rising global populations, increased food demand, as well as demand for dairy and meat-made products, and the intensification of agricultural processes [3,4].

The rise of modern technologies, which not only make farming sustainable but also boost output, food safety, and security while lowering GHG emissions, is the silver lining in this current situation [5–7]. There has been a trend in recent years toward lowering GHG production in the food production sector, but extensive efforts should be made in this direction to uphold global climatic assurances. The global warming potential (GWP) of methane is 25 times more than that of CO_2 over a period of 100 years. The conversion of microbial nitrogen (N) in soil and manure, as well as the dung and urine left behind by grazing animals, is the main source of nitrous oxide. Over a 100-year timeline, nitrous oxide has greater GWP (298 times) than CO_2. About 37% of all agricultural emissions in Europe come from agricultural soils, mostly as a result of synthetic N fertilizers and animal dung in the soil [8,9]. An effective long-term method for reducing climate change is the active management of cultivated soils using the right technologies and agronomic practices. There is overwhelming evidence that, in areas where PA is extensively practiced, water and fertilizer use can be reduced by 20 to 40% without affecting yields, and in some cases can even increase output [10–12].

Precision agriculture technologies (PAT) consider the field's spatial and temporal changes to maximize the usage of agricultural inputs (e.g., irrigation, fuel, and fertilizers). Currently, advanced technologies, e.g., (1) variable rate of nutrient application (VRNT), (2) variable rate of irrigation (VRI), (3) variable rate pesticide application (VRPA), (4) machine guidance (MG), (5) precision physical weeding (PPW), and (6) variable rate planting/seeding (VRP/VRS) are being used in developed countries to boost agricultural production. These variable rate advanced technologies can maintain or increase productivity while lowering GHG emissions from agricultural activities. Policymakers can evaluate the utility of incorporating PA into upcoming agriculture and climate policy tools by examining the role that these tools play in lowering GHG releases and raising farm output. This study focuses on evaluating precision water application techniques, which are suitable for reducing GHG emissions and enhancing overall farm productivity.

2. Greenhouse Gas Emissions in Agriculture Sector

The greenhouse gas emissions from the agriculture sector are relatively high and have a significant impact on climate change and global warming. Significant levels of non-CO_2 emissions, such as methane and nitrous oxide, are released by agricultural operations from crop production and the raising of livestock. The sources and processes of GHG involved in the agriculture sector are shown in Figure 1. According to the US Environmental Protection Agency, in 2020, the agriculture sector accounted for approximately 11% of GHG emissions compared to other sectors, e.g., commercial and residential—13%, land use and forestry—13%, industry—24%, electricity production—25%, and transportation—27% [5,13,14].

As bacterial activity increased under anaerobic conditions with irrigation, more CH4 emissions were produced, indicating that irrigation techniques can have a significant impact on GHG emissions. Additionally, variations in soil moisture have an impact on the redox potential of the soil, which has a substantial impact on the rates of soil GHG emissions [3,10,15]. Therefore, irrigation practices need to be modified and the amount of water required to irrigate the crops must be scheduled according to the crop water requirement to lower GHG emissions. VRI technology is an option that fulfills the spatio-temporal water demands of the crops.

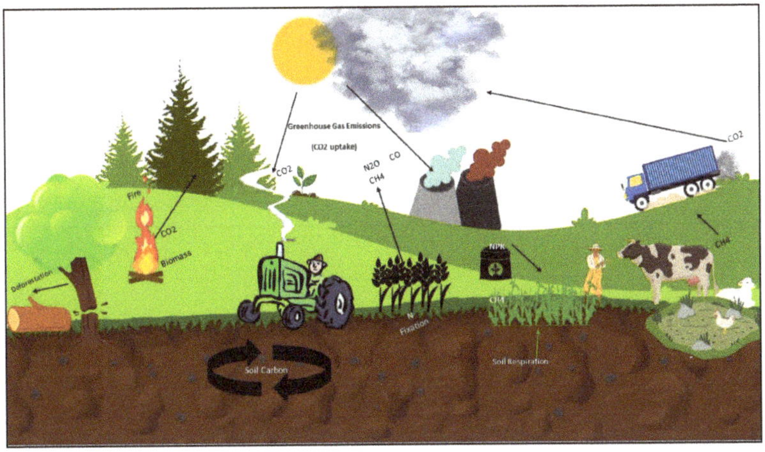

Figure 1. Greenhouse gas emissions in the agriculture sector (Picture credit: Saddam Hussain and Muhammad Habib Ullah).

3. Role of VRI in lowering GHG Emissions

According to a recent study, drip irrigation can significantly lower GHG emissions from soil and sustain the quality of the air without compromising forage crop production. Moreover, the amount of irrigation and nitrogen fertilization is positively correlated with N_2O emissions [1,5,13]. Andrews et al. (2022) [9] found that emissions of CO_2, N_2O, and NO are reduced by up to 62% through subsurface drip irrigation. When N fertilization is paired with accurate precipitation forecasting or scheduled irrigation, GHG emissions can be further decreased. Therefore, a drip irrigation system reduces GHG emissions more than a flood irrigation system. Moreover, the crop water productivity can also be enhanced by adopting an optimum irrigation schedule [4]. If the irrigation method, duration, and amount are modified according to their needs, it can further significantly reduce the GHG emissions.

An irrigation system can optimize irrigation application with the use of cutting-edge technology known as VRI. Since most fields are not uniform, when water is applied evenly, certain portions of the field can be overwatered, while others might stay too dry. This not only affects yield but also alters the GHG emissions cycle. The VRI technology has the potential to reduce over-watering, under-watering, and runoff, ultimately improving soil health and sustains the ecosystem. VRI's contribution to the reduction in greenhouse gas emissions lies in the proper utilization of water, thus reducing the need for energy through pumping. Moreover, proper irrigation schemes prevent extreme soil moisture utilization that increases N_2O emissions. The irrigation water can be saved by around 8–20% using the VRI technology. Many other studies also reported a reduction in water use and an improvement in irrigation efficiency using VRI technology [1,5,9,13]. Hence, the GHG emissions are linked to soil water availability, which depends very much on the amount, timing, and method of irrigation.

4. Conclusion

Irrigation plays an integral role in crop growth, health, and productivity. Precision water management can help to reduce GHG emissions and mitigate climate change. The VRI technology distributes the right amount water to plants and at the right intervals to satisfy the crop water requirement. In addition, less water needed for irrigation requires less pumping energy, powered by either fossil fuel or electricity, indirectly impacting greenhouse gas emissions. This study concludes that a suitable irrigation scheduling and irrigation technology can be used, i.e., VRI reduces GHG emissions. The VRI technique boosts the yield of grains and irrigation efficiency and reduces GHG emissions.

Author Contributions: Conceptualization, S.H., M.J.M.C., M.S.W. and M.H.U.; validation, S.H., and M.J.M.C.; data curation, S.H., M.S.K. and M.H.U.; writing—original draft preparation, S.H, M.S.W. and R.R..; writing—review and editing, S.H., M.S.W., S.R.S., R.R. M.S.K. and M.H.U.; visualization, S.H. and M.S.W.; supervision, M.J.M.C., S.R.S. and M.S.W. All authors have read and agreed to the published version of the manuscript.

Funding: This research received no external funding.

Institutional Review Board Statement: Not applicable.

Informed Consent Statement: Not applicable.

Data Availability Statement: Not applicable.

Acknowledgments: The authors are grateful to the conference organizing committee for providing this wonderful opportunity for conference proceeding publications.

Conflicts of Interest: The authors declare no conflict of interest.

References

1. Montzka, S.A.; Dlugokencky, E.J.; Butler, J.H.J.N. Non-CO2 greenhouse gases and climate change. *Nature* **2011**, *476*, 43–50. [CrossRef] [PubMed]
2. Iqbala, M.M.; Akramb, M.M.; Ahmada, M.; Hussainc, S.; Usmana, G.J.B.D.I.W.R.E. Regional Climatic Response To Global Warming And Agriculture In Pakistan. *Big Data Water Resour. Eng. (BDWR)* **2021**, *2*, 18–23. [CrossRef]
3. MacLeod, M.; Eory, V.; Gruère, G.; Lankoski, J. Cost-effectiveness of greenhouse gas mitigation measures for agriculture: A literature review. In *OECD Food, Agriculture and Fisheries Papers*; OECD: Paris, France, 2015.
4. Waqas, M.S.; Cheema, M.J.M.; Hussain, S.; Ullah, M.K.; Iqbal, M.M.J.A.W.M. Delayed irrigation: An approach to enhance crop water productivity and to investigate its effects on potato yield and growth parameters. *Agric. Water Manag.* **2021**, *245*, 106576. [CrossRef]
5. Kumar, A.; Nayak, A.; Mohanty, S.; Das, B.J.A. Greenhouse gas emission from direct seeded paddy fields under different soil water potentials in Eastern India. *Agric. Ecosyst. Environ.* **2016**, *228*, 111–123. [CrossRef]
6. Aslam, M.; Arshad, M.; Hussain, S.; Usman, M.; Zahid, M.B.; Sattar, J.; Arshad, A.; Iqbal, M.M.; Waqas, M.S.J.P.J.A.S. An integrated approach for estimation of van genuchten model parameters in undisturbed and unsaturated soils. *Pak. J. Agri. Sci.* **2021**, *58*, 1887–1893.
7. Rizvi, F.F.; Ahmed, B.; Hussain, S.; Khan, A.; Raza, M.; Shahid, M.J.I.J.o.G.W. Assessment of climate extremes from historical data (1960-2013) of Soan River Basin in Pakistan. *Int. J. Glob. Warm.* **2021**, *25*, 1–37. [CrossRef]
8. Afroj, M.; Kazal, M.M.H.; Rahman, M.M. Precision agriculture in the World and its prospect in Bangladesh. *Res. Agric. Livest. Fish.* **2016**, *3*, 1–14. [CrossRef]
9. Andrews, H.M.; Homyak, P.M.; Oikawa, P.Y.; Wang, J.; Jenerette, G.D.J.A. Water-conscious management strategies reduce per-yield irrigation and soil emissions of CO2, N2O, and NO in high-temperature forage cropping systems. *Agric. Ecosyst. Environ.* **2022**, *332*, 107944. [CrossRef]
10. Arora, N.K.; Mishra, I.J.E.S. Current scenario and future directions for sustainable development goal 2: A roadmap to zero hunger. *Environ. Sustain.* **2022**, *5*, 1–5. [CrossRef]
11. Aziz, S.; Chowdhury, S.A.J.E. Analysis of agricultural greenhouse gas emissions using the STIRPAT model: A case study of Bangladesh. *Environ. Dev. Sustain.* **2022**, 1–21. [CrossRef] [PubMed]
12. Cheng, H.; Shu, K.; Zhu, T.; Wang, L.; Liu, X.; Cai, W.; Qi, Z.; Feng, S.J.J.o.C.P. Effects of alternate wetting and drying irrigation on yield, water and nitrogen use, and greenhouse gas emissions in rice paddy fields. *J. Clean. Prod.* **2022**, *349*, 131487. [CrossRef]
13. Dilawar, A.; Chen, B.; Ashraf, A.; Alphonse, K.; Hussain, Y.; Ali, S.; Jinghong, J.; Shafeeque, M.; Boyang, S.; Sun, X.; et al. Development of a GIS based hazard, exposure, and vulnerability analyzing method for monitoring drought risk at Karachi, Pakistan. *Geomat. Nat. Hazards Risk* **2022**, *13*, 1700–1720. [CrossRef]
14. Mehmood, F.; Wang, G.; Gao, Y.; Liang, Y.; Zain, M.; Rahman, S.U.; Duan, A.J.W. Impacts of irrigation managements on soil CO2 emission and soil CH4 uptake of winter wheat field in the North China plain. *Water* **2021**, *13*, 2052. [CrossRef]
15. Waqas, M.S.; Cheema, M.J.M.; Waqas, A.; Hussain, S. Enhancing water productivity of potato (Solanum tuberosum L.) through drip irrigation system. In Proceedings of the 2nd International Conference on Horticultural Sciences, Faisalabad, Pakistan, 16–18 February 2016; pp. 18–20.

environmental sciences proceedings

MDPI

Proceeding Paper

Kiwi Plant Growth Monitoring with Soil and Climatic Conditions in the Semi-Arid Region of Pakistan †

Muhammad Umair Arif Khan [1], Muhammad Azam Khan [1,*], Usman Ali Abbasi [1], Muhammad Amin [2], Tahira Kalsoom [1], Abdul Basit [3] and Basit Shahzad [1]

1 Department of Horticulture, PMAS-Arid Agriculture University, Rawalpindi 46000, Pakistan
2 Institute of Geo-Information & Earth Observation, PMAS-Arid Agriculture University, Rawalpindi 46000, Pakistan
3 Green AI, Center of Precision Agriculture, PMAS-Arid Agriculture University, Rawalpindi 46000, Pakistan
* Correspondence: drazam1980@uaar.edu.pk; Tel.: +92-300-6884470

† Presented at the 1st International Precision Agriculture Pakistan Conference 2022 (PAPC 2022)—Change the Culture of Agriculture, Rawalpindi, Pakistan, 22–24 September 2022.

Abstract: Crop growth and yield are influenced by the genetic potential of the cultivar, soil, weather, and cultivation practices, i.e., sowing date, irrigation and fertilizer amount, and biotic stresses. Temporal variation in yield and growth has been largely forecasted using climate as a predictor, which can be achieved by using either an empirical or crop simulation approach for a given location. Climate and soil data collected over agricultural land regularly aid in crop growth monitoring, as well as crop vitality assessment. Crop simulation models (CSM) that have been successful in field-scale applications are now being implemented in GIS framework to simulate and monitor crop growth with remote sensing inputs, allowing for sensitive evaluations of seasonal weather conditions, local variability, and crop management signals. This research was designed to monitor the growth of three varieties of kiwifruit, i.e., Hayward grafted, Green-flesh, and Hayward, in four different localities: Hazaro (Attock), Simli Dam (Pind Begwal), GPU (germ plasm unit) Arid Agriculture University Rawalpindi and ZTBL (Zarai Taraqiati Bank Limited) Farm Islamabad, each of which has different soil and weather conditions. Soil proximal sensors were used to measure soil characteristics, and data loggers were installed in each field to monitor the weather parameters to collect data that influences crops. In this study, we used a quantitative method and GIS-integrated data to assess the impact of soil and climate on kiwifruit growth. It can help policy makers and researchers to identify new agro-climatic zones in Pakistan's semi-arid regions for kiwifruit farming based on this data. In this study, we found that kiwi is very susceptible to temperatures above 40 °C, which cause mortality in kiwifruits plants. Morphological data with respect to soil and climate results showed that green-fleshed and Hayward varieties performed slightly better than Hayward grafted, which was most susceptible to diseases and heat damage.

Keywords: crop simulation models; crop monitoring; remote sensing; soil and climate

Citation: Khan, M.U.A.; Khan, M.A.; Abbasi, U.A.; Amin, M.; Kalsoom, T.; Basit, A.; Shahzad, B. Kiwi Plant Growth Monitoring with Soil and Climatic Conditions in the Semi-Arid Region of Pakistan. *Environ. Sci. Proc.* **2022**, 23, 36. https://doi.org/10.3390/environsciproc2022023036

Academic Editor: Shoaib Rashid Saleem

Published: 17 January 2023

1. Introduction

Kiwifruit is a deciduous perennial climber plant that originated from eastern and southern Asia [1]. Kiwifruit seeds and peels are an excellent source of beneficial compounds, including carotenoids, terpenoids, and polyphenols, and have anti-inflammatory, antimicrobial, antidiabetic, and antioxidant properties; therefore, this fruit has excellent industrial and pharmaceutical potential [2]. Commercial cultivation of kiwifruit started in New Zealand and later spread all over the world [3]. This plant is susceptible to extreme temperatures and frost and requires an optimum temperature ranging from 22 °C to 30 °C for better fruit production [4]. A total of 70 varieties of kiwifruit are grown worldwide, among which only green and golden kiwifruit are used for commercial purposes [5]. In 2020, China had the highest kiwifruit exports, valued at USD 454 million [6].

Approximately 4.41 million metric tons of kiwifruit were produced worldwide in 2020—an increase from the 1.87 million metric tons produced in 2000. In 2020, China and New Zealand produced 2.2 and 0.6 million metric tons, respectively [5]. Commercial kiwi cultivars, including Arctic, Abbott, Allison, Bruno, Hayward, Monty, and Tomuri, are significant globally and are imported into Pakistan from China and other producers. The Pakistan Agricultural Research Council has been testing the Hayward type of kiwi plants in the Hazara belt in past few years. The fruit has also been grown in Battagram, Abbottabad, and Havalian [7].

In Pakistan, kiwi is an exotic plant that is not cultivated at a commercial level. However, research is being conducted to determine suitable climatic zones for fruit production. Climate and soil variability are the two most important factors in plant growth and are essential for determining suitable climatic zones for fruit crop production. Soil characteristics and climatic suitability are the primary factors affecting optimal plant growth and fruit quality [8–10].

The growth cycle of kiwifruit (including budburst, blooming, and fruit maturity) is affected by environmental and soil conditions [9], which determine the occurrence of phenological phases, i.e., lower temperatures in late autumn and winter cease dormancy. Rainfall affects fruit development, and cooler mean temperatures in the fall seem to boost brix. Accordingly, warmer temperatures accelerate phenological phases from budburst through early fruit development, whereas cooler mean temperatures encourage faster maturation rates in autumn and the end of dormancy in winter. In this way, kiwifruit vine is sensitive to frost and high temperatures and cannot withstand harsh weather [2,11].

Soil nutrient status also affects plant productivity, as excess or deficient N, P, and K can represent potential risks, reducing fruit quality, quantity, and plant growth [12,13]. Like soil nutrient status, pH and electric conductivity can also affect kiwifruit crop productivity, as high pH can significantly constrain nutrient availability [14].

As the world population is increasing daily and modern technologies are progressing rapidly, agriculture interventions to monitor crop growth need time to fulfill nutritional requirements. In the current research, GIS and sensor-based approaches were applied to monitor the growth of kiwifruit, considering climate and soil parameters [15,16]. It is a well-established practice to employ GIS and RS data for crop monitoring during all stages of activity, i.e., planning, analysis, and output [17]. For this purpose, different sensors are used to determine fertilizer requirements, water availability, and pest infestations. Precision agriculture provides explicit real-time estimations through satellite systems. This technology provides accurate field data [18], reduces labor costs and input resources, and boosts agricultural productivity [17,19,20].

It is essential to study the two crucial factors of climate and soil to monitor the growth of kiwifruit. Using proximal sensors, in this study, we intended to evaluate the morphological response of kiwifruit in a semi-arid region, determine the influence of climate and soil variability on kiwifruit, and compare the growth of different kiwifruit varieties in terms of climate and soil suitability.

2. Materials and Methods

2.1. Experimental Site

The research experiment was carried out at 4 different locations in the semi-arid region of Potohar, Pakistan. These locations included Simli Dam (Pind Begwal; 33°41′57″ N 73°15′52″ E), Zarai Taraqiati Bank Limited (Farm Islamabad; 13°39′59″ N 73°06′01″ E), Attock (Tehsil Hazaro; 33°56′55″ N 72°24′52″ E), and the Germ Plasm Unit (Pir Mehr Ali Shah Arid Agriculture University, Rawalpindi; 13°38′48″ N 73°04′50″ E). Three varieties were selected for this research, i.e., green-fleshed, Hayward grafted, and Hayward. Satellite imagery of the above-mentioned locations via GPS logger are illustrated in Figure 1:

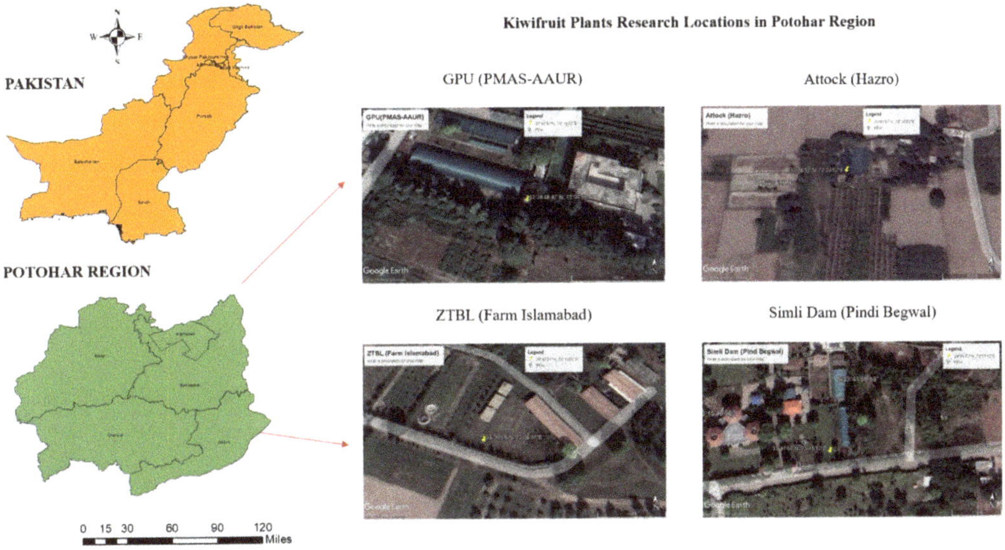

Figure 1. Kiwifruit plant research locations in Pothohar region.

2.2. Study Parameters

Morphological growth parameters include the number of leaves, number of shoots, stem diameter (cm), plant height (cm), inter-nodal distance (cm), and leaf area (cm). To monitor the growth of kiwifruit plants, soil (NPK concentration, pH, electrical conductivity, and organic matter) and climate (temperature, rainfall, and relative humidity) parameters were measured. Soil and climate data were utilized to estimate agroclimatic appropriateness their effect on plant morphological growth.

2.3. Monitoring Parameters

Rainfall data were collected from the Pakistan Meteorological Department. Temperature data were collected using a data logger installed at the research locations. Soil parameter results from the lab and proximal sensors were compared. A calibrated scale was used to assess morphological growth from March 2022 to August 2022.

2.4. Data Evaluation and Analysis

RCBD analysis was used to evaluate data. By considering the soil and climatic parameters, varieties for a specific location were planned for the next growing season, and the influence of these parameters on morphological growth were also considered.

3. Results and Discussion

The chemical properties of the soil used for experiments at the four locations are shown in Table 1. This test indicated the nutrient status of soil in which kiwifruit plants were grown.

Temperature data were collected through a data logger, and rainfall data were collected from the Pakistan Meteorological Department, as shown in Table 2. These data, which were observed during a six-month period from March to August, show the average rainfall, minimum temperature, and maximum temperature during each month.

Table 1. Average data of chemical properties and nutrient status of soil collected through proximate soil sensors and soil fertility tests.

	Attock	Simly	GPU	ZTBL
Nitrogen (mg/kg)	14	16	16	12
Phosphorous (ppm)	0.919	0.615	1.046	0.288
Potassium (ppm)	3.5	1.3	3.9	2.1
EC (us/cm)	0.4	0.1	0.1	0.1
pH	6.4	6.19	6	6.5
Organic matter (%)	0.93	1.28	0.24	0.59

Table 2. Climatic factors recorded at different locations.

Location	Month	Climatic Factors				
		T °C (Average Min)	T °C (Min)	T °C (Average Max)	T °C (Max)	Rainfall (mm)
Attock	March	20	8.1	36	37	66.3
	April	21.2	17.2	37.2	38	50.7
	May	24.6	19.5	40	43	33
	June	31	22.5	45	49	32.7
	July	31.5	24.3	45.5	47	99.5
	August	29.5	23.1	43	46	97.1
ZTBL	March	17	7.1	33	34.5	73.8
	April	18	14.3	34.3	35.1	59.7
	May	21.5	16.1	37	40	39.2
	June	28.2	19.5	42.3	46	62.2
	July	27.5	19.8	42	44	267
	August	22.9	20	39.2	42	309.9
Simly	March	16.5	7.0	32	33.5	140.39
	April	17.2	13.2	33.5	34.5	94.23
	May	20.8	14.8	36	39.1	55.02
	June	27.5	18.1	41.2	43.5	63.55
	July	26.6	19.3	41	43.1	261.5
	August	21.5	19.5	38.5	41	240.31
GPU	March	18	7.5	34	35	71.8
	April	19.6	15.5	35.1	36	57.7
	May	22	17.5	38	41	30
	June	29	20.2	43.2	47.3	53.3
	July	29.5	21.1	43.3	45.2	237
	August	27.5	20.3	41.8	43.8	236

As shown in Table 3, different growth patterns were observed among the three varieties at each location. For each variety, stem diameter was almost constant at every location, but the highest stem diameter was recorded (0.85 cm) in the green-fleshed variety at Simly Dam (Pind Begwal). Minute variation in stem diameter is due to variations in soil chemical composition. An increases in stem diameter often followed canopy responses and soil chemical composition, as described in [7]. Collected data show a major difference in the plant height of each variety at different locations. This variation is due to weather conditions; rainfall of more than 150 mm in June and July caused water flooding in the root zone, affecting the root zone's oxygen ratio. In terms of plant height, the Hayward kiwi plant reached a height of 126.4 cm in Attock, followed by 125.32 cm in GPU, 125.67 cm in ZTBL, and 75.5 cm in the Simly Dam kiwifruit orchard. The height of the Hayward grafted variety was 128.83 cm at the Simly Dam location. The third variety, green-fleshed, was most heightened at each location as compared to other varieties, with a maximum height of 246.17 cm at the Simly Dam location, followed by 228.17 cm at ZTBL, 219.3 cm at GPU, and 173.35 cm at Attock. Tree growth and development are determined by the plant's height in the local climate [21]. The genetic makeup of the variety and the impact of environmental factors lead to diversity in plant height among olive types, which behave differently in various climatic situations. The efficient use of plant nutrients and water at various locations is the cause of the variance in plant height in kiwifruit [12].

Table 3. Growth parameter analysis of three varieties (Hayward, Hayward grafted, and green-fleshed) at Attock, GPU, ZTBL, and Simly Dam.

Variety	Location	Parameters						
		Stem Dia. (cm)	Plant Height (cm)	Internodal Distance (cm)	No. of Shoots	No. of Leaves	Disease (%)	Heat Damage (%)
Hayward	Attock	0.755	126.4	8.625	10	50.167	2.933	23.833
	GPU	0.5867	125.32	8.6267	9.8333	50.5	4.6667	5.5
	ZTBL	0.5467	125.67	8.63	12	52	5	8.833
	Simly Dam	0.8333	75.5	4.8167	10.5	46.667	3.8333	3.3333
Hayward grafted	Attock	0.6083	94.2	5.7167	5.667	54	7.133	25.633
	GPU	0.5583	101.67	6.6	8.3333	52	5.8333	10
	ZTBL	0.515	111.67	8.6267	11.667	52.5	5.3333	12.167
	Simly Dam	0.8367	128.83	6.5133	10.167	64.833	4.5	4.8333
Green-fleshed	Attock	0.6283	173.35	7.5167	6.333	53.833	4.555	4.23
	GPU	0.56	219.28	8.15	7.8333	57.333	5.1667	5.167
	ZTBL	0.5133	228.17	8.2167	9	72.833	4.5	5.667
	Simly Dam	0.8533	246.17	8.2167	10.333	82.667	3.5	3.5

The climate data at all locations are shown in Figure 2, this climatic data affects kiwifruit morphological growth as shown in Table 3. The data on climatic factors shown in Table 2 indicate that rainfall and temperature variation affected plant growth, as reflected by internodal distance, the number of shoots, and the number of leaves. The internodal distance reached a maximum value (8.62 cm) in the Hayward grafted variety at ZTBL and a minimum value (4.82 cm) in the Hayward variety at Simly Dam. Internodal distance determines the morphology of the canopy and the porosity of the branches, which favors light and air permeability [22]. All vegetative tissues of plants share metabolites, so their effects on intermodal tissues vary [22,23]. The maximum number of shoots was 11.7 in the Hayward grafted variety, and the maximum number of leaves was 82.7 in the green-fleshed variety. When the internode size of native spearmint tissues is increased, the number of shoots decreases [24].

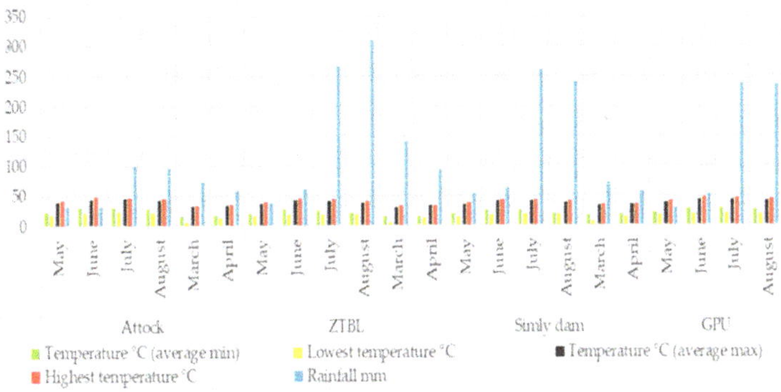

Figure 2. Climatic factors recorded at different locations.

The percentage of disease caused by genera *Sphingobium* and *Phytophthora* was 7.133% in the Hayward grafted variety and 2.93% in the Hayward variety at Attock kiwi orchard. These pathogens are responsible for the loss of kiwifruit vine [25]. In June and July, temperatures above 40 °C adversely affect kiwifruit plants, making them more vulnerable to sudden and extreme weather events, such as very high, quickly rising temperatures and rains, which occur more frequently in the summer and cause intense and abrupt water stress [26]. Heat damage percentage was also recorded as highest in the Hayward grafted and lowest in the Hayward and green-fleshed varieties at Simly Dam. These data show that green-fleshed and Hayward varieties are the most suited to the local climate, and the Hayward grafted variety is the most susceptible to diseases and heat damage (Table 3), which can have long-term effects on the growth of the kiwifruit plant. The physiological responses of the kiwifruit vine to environmental stress can indicate for opportunities that can be overcome through breeding and, cultural techniques. Spatial variation in leaf arrangement and temporal variation in irradiance conditions lead to highly dynamic irradiance of individual leaves within the canopy [26]. Overall, the green-fleshed kiwi variety performed better than the Hayward and Hayward grafted varieties in terms of morphological parameters at all research locations. The climatic conditions at Simly Dam are most suitable for kiwi plants. Because kiwi plants like humidity, they are widely adopted in humid areas of China and New Zealand [27]. Rainfall of more than 150 mm causes flooding in the root zoon of kiwi plant, and the physiology of plants and the anatomical characteristics of the xylem point to a potential involvement of weather. This role might be highly harmful to kiwifruit, which is known as an anisohydric plant [28].

4. Conclusions

Identifying new agro-climatic zones in Pakistan's semi-arid regions for kiwifruit farming is possible. This plant is being introduced in Pakistan to attain significant benefits. For this purpose, a number of kiwifruit varieties were grown in different areas to determine suitable soil and climate conditions for growth and commercial production. The morphological characteristics of three varieties were recorded at four locations, and the effects of climate and soil were observed. Crop simulation models (CSMs) are now being applied in GIS framework to simulate and monitor crop growth with remote sensing inputs, allowing for sensitive evaluation of seasonal weather conditions, local variability, and crop management. In this study, we found that kiwi is very susceptible to warm weather, is sensitive to frost and high temperatures, and cannot bear harsh weather; specifically, temperatures above 40 °C cause mortality in kiwifruit plants. Morphological results show that green-fleshed and Hayward varieties performed slightly better than the Hayward grafted variety, which was found to be the most susceptible to diseases and heat damage, which can have long-term effects on the fruiting of kiwi plants.

Author Contributions: All authors contributed equally in conducting and compiling research experiments and data analysis. M.U.A.K. performed experiments under the supervision and guidance of M.A.K., U.A.A. and M.A., data collection; M.U.A.K., B.S. and T.K.; Investigation; M.A.K., U.A.A., M.A. and U.A.A., A.B. and M.U.A.K., was responsible for writing—review and editing. All authors have read and agreed to the published version of the manuscript.

Funding: This research was funded by Punjab Agriculture Research Board (PARB) project 1085.

Institutional Review Board Statement: Not applicable.

Informed Consent Statement: Not applicable.

Data Availability Statement: The data are available upon request.

Acknowledgments: This research was conducted under the project title Kiwi Plant Growth Monitoring with Soil and Climatic Conditions in the Semi-Arid Region of Pakistan funded by PARB (1085) and was part of an M.Sc. (hons) research thesis.

Conflicts of Interest: The authors declare no conflict of interest. This research was designed and executed by the authors, and the funding agency had no role in the design, collection and interpretation of data, writing of the manuscript, or in the decision to publish the results.

References

1. Ferguson, A.R. Kiwifruit: A botanical review. *Hortic. Rev.* **2011**, *6*, 6.
2. Ferguson, A.R.; Seal, A.G. Kiwifruit. In *Temperate Fruit Crop Breeding*; Springer: Dordrecht, The Netherlands, 2008; pp. 235–264.
3. Kohale, V.S. *Advances in Horticulture Sciences*; Integrated Publications: New Delhi, India, 2022; Volume 6. [CrossRef]
4. Huang, S.; Ding, J.; Deng, D.; Tang, W.; Sun, H.; Liu, D.; Zhang, L.; Niu, X.; Zhang, X.; Meng, M. Draft genome of the kiwifruit Actinidia chinensis. *Nat. Commun.* **2013**, *4*, 1–9. [CrossRef] [PubMed]
5. Xu, S.; Ning, J.; Li, Y.; Zhang, Y.; Xu, G.; Huang, X.; Deng, R.H. Match in my way: Fine-grained bilateral access control for secure cloud-fog computing. *IEEE Trans. Dependable Secur. Comput.* **2020**, *19*, 1064–1077. [CrossRef]
6. Silva, S.S.; Justi, M.; Chagnoleau, J.-B.; Papaiconomou, N.; Fernandez, X.; Santos, S.A.; Passos, H.; Ferreira, A.M.; Coutinho, J.A. Using biobased solvents for the extraction of phenolic compounds from kiwifruit industry waste. *Sep. Purif. Technol.* **2023**, *304*, 122344. [CrossRef]
7. Islam Zeb, M.S.; Ullah, I.; Rahman, J.; Ahmad, Z.; Bibi, F.; Begum, S.K.N.; Ali, S.; Khan, A. Optimization of propagation techniques and timimg for the production of kiwi fruit (Actinidia chinensis) plant at Khyber Pakhtun Khwa Mingora Swat. *Pure Appl. Biol.* **2017**, *6*, 889–896.
8. Yuan, L.; Yanan, T.; Liling, C. Effect of fertilization on kiwifruit yield and quality. *J. Northwest Agric. For. Univ.* **2011**, *39*, 171–176.
9. Zuoping, Z.; Min, D.; Sha, Y.; Zhifeng, L.; Qi, W.; Jing, F.; Yan'an, T. Effects of different fertilizations on fruit quality, yield and soil fertility in field-grown kiwifruit orchard. *Int. J. Agric. Biol. Eng.* **2017**, *10*, 162–171.
10. Waqas, M.S.; Cheema, M.J.M.; Waqas, A.; Hussain, S. Enhancing water productivity of potato (*Solanum tuberosum* L.) through drip irrigation system. In Proceedings of the 2nd International Conference on Horticultural, Faisalabad, Pakistan, 16–18 February 2016; pp. 18–20.
11. Waqas, M.S.; Cheema, M.J.M.; Hussain, S.; Ullah, M.K.; Iqbal, M.M. Delayed irrigation: An approach to enhance crop water productivity and to investigate its effects on potato yield and growth parameters. *Agric. Water Manag.* **2021**, *245*, 106576. [CrossRef]
12. Wang, N.; He, H.; Lacroix, C.; Morris, C.; Liu, Z.; Ma, F. Soil fertility, leaf nutrients and their relationship in kiwifruit orchards of China's central Shaanxi province. *Soil Sci. Plant Nutr.* **2019**, *65*, 369–376. [CrossRef]
13. Hussain, S.; Cheema, M.J.M.; Waqas, M.S.; Waqas, A. Effect of potash application on growth and yield of onion crop with drip system. In Proceedings of the 2nd International Conference on Horticultural, Faisalabad, Pakistan, 16–18 February 2016; pp. 195–201.
14. Dede, G.; Özdemir, S.; Dede, Ö.H.; Altundağ, H.; Dündar, M.; Kızıloğlu, F.T. Effects of biosolid application on soil properties and kiwi fruit nutrient composition on high-pH soil. *Int. J. Environ. Sci. Technol.* **2017**, *14*, 1451–1458. [CrossRef]
15. Kasampalis, D.A.; Alexandridis, T.K.; Deva, C.; Challinor, A.; Moshou, D.; Zalidis, G. Contribution of remote sensing on crop models: A review. *J. Imaging* **2018**, *4*, 52. [CrossRef]
16. Hussain, S.; Malik, S.; Masud Cheema, M.; Ashraf, M.U.; Waqas, M.; Iqbal, M.; Ali, S.; Anjum, L.; Aslam, M.; Afzal, H. An overview on emerging water scarcity challenge in Pakistan, its consumption, causes, impacts and remedial measures. *Big Data Water Resour. Eng.* **2020**, *1*, 22–31. [CrossRef]
17. Shafeeque, M.; Sarwar, A.; Basit, A.; Mohamed, A.Z.; Rasheed, M.W.; Khan, M.U.; Buttar, N.A.; Saddique, N.; Asim, M.I.; Sabir, R.M. Quantifying the Impact of the Billion Tree Afforestation Project (BTAP) on the Water Yield and Sediment Load in the Tarbela Reservoir of Pakistan Using the SWAT Model. *Land* **2022**, *11*, 1650. [CrossRef]
18. Mandal, D.; Kumar, V.; Ratha, D.; Dey, S.; Bhattacharya, A.; Lopez-Sanchez, J.M.; McNairn, H.; Rao, Y.S. Dual polarimetric radar vegetation index for crop growth monitoring using sentinel-1 SAR data. *Remote. Sens. Environ.* **2020**, *247*, 111954. [CrossRef]
19. Supit, I.; Van Diepen, C.; De Wit, A.; Wolf, J.; Kabat, P.; Baruth, B.; Ludwig, F. Assessing climate change effects on European crop yields using the Crop Growth Monitoring System and a weather generator. *Agric. For. Meteorol.* **2012**, *164*, 96–111. [CrossRef]
20. Malik, S.; Hussain, S.; Waqas, M.S. Effect of water quality and different meals on growth of Catla catla and Labeo rohita. *Big Data Water Resour. Eng.* **2020**, *1*, 4–8. [CrossRef]
21. Nafees, M.; Jafar Jaskani, M.; Ahmad, I.; Ashraf, I.; Maqsood, A.; Ahmar, S.; Azam, M.; Hussain, S.; Hanif, A.; Chen, J.-T. Biochemical analysis of organic acids and soluble sugars in wild and cultivated pomegranate germplasm based in Pakistan. *Plants* **2020**, *9*, 493. [CrossRef]
22. Rowan, D.; Boldingh, H.; Cordiner, S.; Cooney, J.; Hedderley, D.; Hewitt, K.; Jensen, D.; Pereira, T.; Trower, T.; McGhie, T. Kiwifruit Metabolomics—An Investigation of within Orchard Metabolite Variability of Two Cultivars of Actinidia chinensis. *Metabolites* **2021**, *11*, 603. [CrossRef]
23. Sumrah, M.A.; Jan, M.; Hussain, A.; Akhtar, S.; Nawaz, H.; Afzal, M.; Umar, H. Evaluation of some promising varieties of olive (*Olea europaea* L.) for growth and yield under Pothwar Regions of Punjab, Pakistan. *Pak. J. Agric. Res* **2021**, *34*, 446–453. [CrossRef]
24. Kim, M.; Kim, S.-C.; Moon, D.-Y.; Song, K.J. Rapid shoot propagation from micro-cross sections of kiwifruit (Actinidia deliciosa cv. 'Hayward'). *J. Plant Biol.* **2007**, *50*, 681–686. [CrossRef]
25. Donati, I.; Cellini, A.; Sangiorgio, D.; Caldera, E.; Sorrenti, G.; Spinelli, F. Pathogens associated to kiwifruit vine decline in Italy. *Agriculture* **2020**, *10*, 119. [CrossRef]
26. Smith, G.S.; Buwalda, J.G. Kiwifruit. In *Handbook of Environmental Physiology of Fruit Crops*; CRC Press: Boca Raton, FL, USA, 2018; pp. 135–163.

27. Scortichini, M.; Marcelletti, S.; Ferrante, P.; Petriccione, M.; Firrao, G. Pseudomonas syringae pv. actinidiae: A re-emerging, multi-faceted, pandemic pathogen. *Mol. Plant Pathol.* **2012**, *13*, 631–640. [CrossRef] [PubMed]
28. Bardi, L.; Nari, L.; Morone, C.; Solomita, M.; Mandalà, C.; Faga, M.G.; Migliori, C.A. Kiwifruit Adaptation to Rising Vapor Pressure Deficit Increases the Risk of Kiwifruit Decline Syndrome Occurrence. *Horticulturae* **2022**, *8*, 906. [CrossRef]

environmental sciences proceedings

MDPI

Proceeding Paper

Variable-Rate Fertilization for Citrus Orchard Management †

Syed Muiz Mehdi Naqvi [1], Muhammad Azam Khan [1,*], Irfan Ali [1], Shoaib Saleem [2,3], Basit Shahzad [1], Abdul Basit [2] and Muhammad Farhan [1]

1 Department of Horticulture, PMAS Arid Agricultural University, Rawalpindi 46000, Pakistan
2 Green AI, Center of Precision Agriculture, PMAS Arid Agriculture University, Rawalpindi 46000, Pakistan
3 Department of Farm Machinery and Precision Engineering, PMAS Arid Agricultural University, Rawalpindi 46000, Pakistan
* Correspondence: drazam1980@uaar.edu.pk; Tel.: +92-3006884470
† Presented at the 1st International Precision Agriculture Pakistan Conference 2022 (PAPC 2022)—Change the Culture of Agriculture, Rawalpindi, Pakistan, 22–24 September 2022.

Abstract: All tropical and subtropical areas of the world are suitable for citrus cultivation. In managing fertilizer application efficiency in orchards, variable-rate technology (VRT) has been demonstrated to be an important element. This article aims to study the significance of variable-rate fertilization for citrus in the arid region of Pakistan. The NPK was calculated before the application of the variable-rate fertilizer. The plant height and stem girth were determined before and after fertilizer application. The preliminary results revealed that the stem girth performed significantly better than the plant height after applying fertilizer by VRT. The preliminary results showed a significant difference in the fruit yield between the VRT and uniform-rate fertilizer application.

Keywords: variable-rate technology; citrus; stem girth; plant height; fertilizers

1. Introduction

Pakistan is a producer and exporter of citrus, ranked sixth overall globally, with an average production of 2,468,671 tons in an area of 181,650 hectares [1]. Citrus fruit occupies a dominant economic position in the world's fruit sector and is commercially grown in over 130 countries. However, Pakistan only contributes 2.9% of the world's production of citrus fruit, such as mandarin and oranges, due to its low average yield compared to worldwide trends [2]. The cost of production has increased due to rising inflation, while the yield has stagnated. Although citrus has the potential to be a significant crop, proper research has not been undertaken to boost its productivity. Despite the introduction of modern technologies, such as precision agriculture, VRF, etc., and their impressive outcomes, farmers have not yet adopted them to improve the yield of their orchards [3]. VRF application has the potential to improve fertilizer use efficiency, reduce the cost of production, and reduce the environmental impacts [4,5]. Variable-rate technology (VRT) is a key site-specific precision agriculture technique that empowers variable dose input control in the field dependent on the crop and soil spatial variability [6,7]. Although this technology is well known and efficient, growers have not yet implemented it as their major method of fertilization [8]. The objective of this study was to reduce the loss of fertilizers via the traditional method of fertilization and to reduce the toxicity caused by the excessive use of fertilizers.

Citation: Naqvi, S.M.M.; Khan, M.A.; Ali, I.; Saleem, S.; Shahzad, B.; Basit, A.; Farhan, M. Variable-Rate Fertilization for Citrus Orchard Management. *Environ. Sci. Proc.* **2022**, 23, 37. https://doi.org/10.3390/environsciproc2022023037

Academic Editor: Saddam Hussain

Published: 1 February 2023

2. Materials and Methods

2.1. Experimental Site

Citrus groves located at the University Research Farm Koont of PMAS-AAUR, Rawalpindi, were selected for this research to evaluate the impact of the VRT on citrus growth and yield; the map of the area is shown in Figure 1.

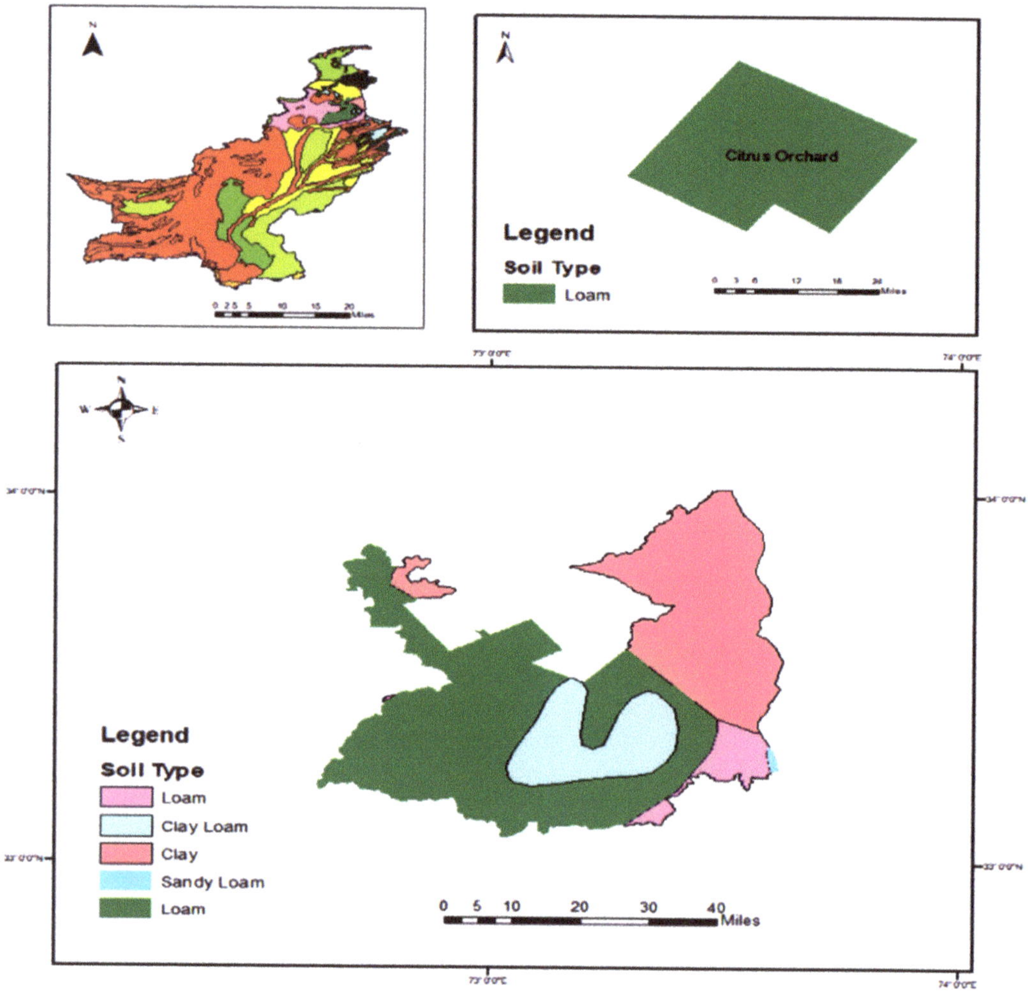

Figure 1. Map of the study area, the citrus orchard, and Rawalpindi.

2.2. Experiment Framework

The selected field for the experiment had been under URF for the past few years. Plants with an age of more than five years were selected for this research. The VRF was applied based on the canopy volume. The plants were divided into three groups having different canopy volumes, as shown in Table 1. The plant canopy was measured with the formula $1/2\pi r^2 \times h$.

Table 1. Plant groups based on canopy volume.

Groups ID	Canopy Volume (ft^3)	Nitrogen (g)	Phosphorous (g)	Potassium (g)
A1	300–450	875	1390	500
A2	450–600	1050	1675	600
A3	600–750	1250	1940	700

Soil sampling was performed to verify the accuracy of the recommended doses, carried out by soil proximal sensors. The plant height, stem girth, and fruit yield were measured to evaluate the effect of the VRF. A statistical comparison was performed to evaluate the impacts of the variable-rate fertilization on the plant height and stem girth.

3. Results

3.1. Plant Height

The plant height plays a significant role in the development of plant physiology [4]. Plant height readings were taken at regular intervals over time. The results indicated that the heights of the plants under the variable-rate fertilization (VRF) and uniform-rate fertilization showed only a small difference; the results are shown in Table 2. On the other hand, the plants under VRF performed well with less input as compared to the plants under uniform-rate fertilization.

Table 2. Comparison of the plant height under the URF and VRF (2021).

Canopy Volume (ft^3)	Sample	N	Mean	*p*-Value	Mean	**p*-Value
			July 2021			July 2022
300–450	VRF	3	8.20	0.0503	8.63	0.018087
	URF	3	8.03		8.60	
450–600	VRF	3	10.33	0.0236	10.83	0.008688
	URF	3	9.83		10.43	
600–750	VRF	3	9.90	0.0230	10.50	0.158950
	URF	3	10.73		11.43	

****p*-value** less than 0.05 shows significant results.

3.2. Stem Girth

A significant influence was observed in the stem girth in the plants receiving VRF as compared to the plants under URF, especially in the second interval of the reading; the results are shown in Table 3.

Table 3. Analysis of variance of the stem girth at 50% (2021).

Canopy Volume (ft^3)	Sample	N	Mean	*p*-Value	Mean	**p*-Value
			July 2021			July 2022
300–450	VRF	3	15.66	0.400985	17.10	0.116486
	URF	3	15.33		0.52	
450–600	VRF	3	20.17	0.095855	1.60	0.113882
	URF	3	19.00		3.00	
600–750	VRF	3	18.50	0.044226	1.25	0.263621
	URF	3	18.67		2.12	

****p*-value** less than 0.05 shows significant results.

3.3. Interpolation of the Soil Nutrients

In Figure 2, the maps show the current nutrient status of the nitrogen, phosphorous, and potassium in the citrus orchards after the application of fertilizers with the VRT methods. Soil data were taken from 15 different stations by soil proximal sensors after the application of fertilizer with the VRT method. The maps show that the soil contained a reliable concentration of all primary nutrients.

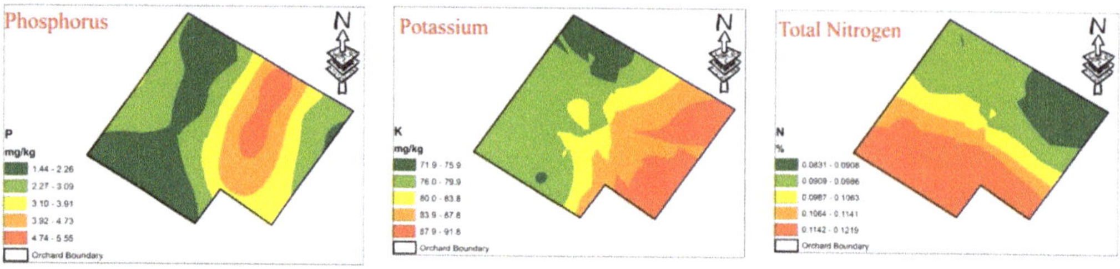

(**a**) (**b**) (**c**)

Figure 2. Kriged interpolated maps of (**a**) potassium, (**b**) nitrogen, and (**c**) phosphorous.

3.4. Fruit Yield

It is essential to note that the impact of the VRF and URF on citrus yield depends on a variety of considerations, including the type of citrus being cultivated and the soil conditions. The fruit yield of the plants under the VRF treatment was higher than the plants under the URF treatment, as shown in Figure 3. Moreover, there were fewer disease attacks on the fruits after maturity in the plants under the VRF treatment.

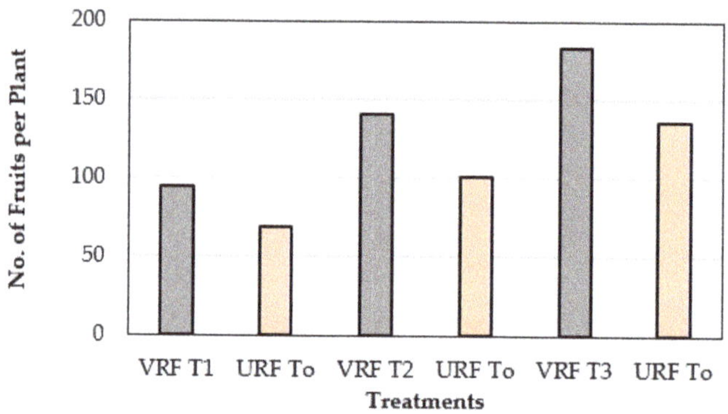

Figure 3. Number of fruits per plant vs. treatments based on the canopy volume.

4. Conclusions

A calculated amount of fertilizer was applied to the plants under the VRF treatment as compared to the plants under the uniform-rate fertilization (URF) treatment. The random or excessive use of fertilizers increases the risk of soil toxicity. It is also considered an unsustainable practice. The preliminary results showed that the VRT method reduced the fertilizer quantity, which also reduced the cost of fertilization for the farmer. Due to the abovementioned benefits, VRT has the potential to increase the production of crops as well as their quality by enhancing their physiology.

Author Contributions: Conceptualization, methodology, software, validation, S.M.M.N., M.A.K., I.A. and A.B.; formal analysis, S.S., M.F. and B.S.; investigation, S.S.; resources, I.A. and M.A.K.; data curation, A.B. and S.S.; writing—original draft preparation, S.M.M.N.; writing—review and editing, S.S., A.B., B.S. and M.A.K.; visualization, A.B.; supervision, M.A.K. and I.A.; project administration, M.A.K. and S.S. All authors have read and agreed to the published version of the manuscript.

Funding: This research received no external funding.

Institutional Review Board Statement: Not applicable.

Informed Consent Statement: Not applicable.

Data Availability Statement: The data can be made available upon request.

Conflicts of Interest: The authors declare no conflict of interest.

References

1. Naz, S.; Shahzadi, K.; Rashid, S.; Saleem, F.; Zafarullah, A.; Ahmad, S. Molecular characterization and phylogenetic relationship of different citrus varieties of Pakistan. *JAPS J. Anim. Plant Sci.* **2014**, *24*, 315–320.
2. Hassan, R.A.; Ali, S.; Zaheer, M.S.; Ali, H.H.; Iqbal, J.; Habib, A.; Nadeem, M.A.; Mumtaz, M.Z. In-vitro and in-vivo evaluation of different fungicides against leaf blight causing fungus Alternaria cucumerina in bitter gourd. *J. Saudi Soc. Agric. Sci.* **2022**, *21*, 208–215. [CrossRef]
3. Seelan, S.K.; Laguette, S.; Casady, G.M.; Seielstad, G.A. Remote sensing applications for precision agriculture: A learning community approach. *Remote. Sens. Environ.* **2003**, *88*, 157–169. [CrossRef]
4. El-Gioushy, S. Productivity, fruit quality and nutritional status of Washington navel orange trees as influenced by foliar application with salicylic acid and potassium silicate combinations. *J. Hortic. Sci. Ornam. Plants* **2016**, *8*, 98–107.
5. Yang, C.; Everitt, J.; Bradford, J. Comparisons of uniform and variable rate nitrogen and phosphorus fertilizer applications for grain sorghum. *Trans. ASAE* **2001**, *44*, 201. [CrossRef]
6. Colaço, A.; Molin, J.P. Variable rate fertilization in citrus: A long term study. *Precis. Agric.* **2017**, *18*, 169–191. [CrossRef]
7. Ehsani, R.; Schumann, A.; Salyani, M. *Variable Rate Technology for Florida Citrus: AE444/AE444*; University of Florida: Gainesville, FL, USA, 2009.
8. Nawar, S.; Corstanje, R.; Halcro, G.; Mulla, D.; Mouazen, A.M. Delineation of soil management zones for variable-rate fertilization: A review. *Adv. Agron.* **2017**, *143*, 175–245.

environmental sciences proceedings

MDPI

Proceeding Paper

Application of Sensor-Based Precision Irrigation Methods for Improving Water Use Efficiency of Maize Crop †

Muhammad Abubakar Aslam [1,2,*], Muhammad Jehanzeb Masud Cheema [2,3], Shoaib Saleem [4,5], Abdul Basit [1,5], Saddam Hussain [1,2] and Muhammad Sohail Waqas [6]

1 Department of Irrigation and Drainage, University of Agriculture Faisalabad, Faisalabad 38000, Pakistan
2 National Center of Industrial Biotechnology, PMAS Arid Agriculture University, Rawalpindi 46000, Pakistan
3 Faculty of Agricultural Engineering and Technology, PMAS Arid Agricultural University, Rawalpindi 46000, Pakistan
4 Department of Farm Machinery and Precision Engineering, Faculty of Agricultural Engineering and Technology, PMAS Arid Agricultural University, Rawalpindi 46000, Pakistan
5 Green AI, Center of Precision Agriculture, PMAS Arid Agriculture University, Rawalpindi 46000, Pakistan
6 Soil Conservation Group, Agriculture Department (Field Wing), Government of the Punjab, Rawalpindi 46000, Pakistan
* Correspondence: engrabubakar@hotmail.com; Tel.: +92-322-6006200

† Presented at the 1st International Precision Agriculture Pakistan Conference 2022 (PAPC 2022)—Change the Culture of Agriculture, Rawalpindi, Pakistan, 22–24 September 2022.

Abstract: Soil moisture sensors and hydraulic modeling play a vital role in managing surface irrigation systems. Crop water productivity can be improved by managing the inflow cut-off time and optimizing the other field scale measurements. As such, hydraulic modelling and field experiments were carried out at the University of Agriculture Faisalabad-Pakistan. The soil moisture sensor (SEN-13322) and the WinSRFR model were used for this purpose. In total, nineteen treatments including eighteen simulated treatments and one conventional treatment were designed at two levels of discharge (Q_1:0.0025 and Q_2:0.0035 $m^3 s^{-1}$), at three sensor positions (S_1:55%, S_2:65%, and S_3:75%) across the field length, as well as with three different border widths (B_1:6.4m, B_2:8.5m, and B_3:10.7m) after successful sensor and model calibration during the two growing seasons of 2016–2017 and 2017–2018. The results revealed a significant difference between the means and the treatment T_{10} i.e., $Q_2S_1B_1$ that were found to be highly efficient and uniform.

Keywords: cut-off time; WinSRFR model; border width; irrigation science; crop production

Citation: Aslam, M.A.; Cheema, M.J.M.; Saleem, S.; Basit, A.; Hussain, S.; Waqas, M.S. Application of Sensor-Based Precision Irrigation Methods for Improving Water Use Efficiency of Maize Crop. *Environ. Sci. Proc.* **2022**, *23*, 38. https://doi.org/10.3390/environsciproc2022023038

Academic Editor: Tahir Iqbal

Published: 1 February 2023

1. Introduction

Water shortage has become a major problem and challenge around the world in recent decades. The shortage of water has also affected agricultural output and encouraged scientists to consider how best to manage the available water resources. Pakistan is a developing nation that struggles with problems, including the lack of water as a result of industrialization, urbanization, and the rising demand for water [1]. The current per capita amount of water resources in Pakistan ranges from 5600 m^3 to 1000 m^3 [2]. To address the issue of water scarcity and to satisfy the demand for agricultural products while maximizing the use of limited amounts of available water resources and minimizing water losses, efficient irrigation methods at the farm level are needed [3]. The farmers in Pakistan irrigate their crops using age-old techniques (i.e., flood irrigation) that involve flowing water like a sheet along the field's length.

With the flood irrigation technique, uncontrolled water runs toward the end of the field, which usually waterlogs the crop and irrigates it more than the crop water requirement, resulting in overall yield decrease and the wastage of water. If irrigation water is cut off at the right time before it reaches the field tail-end boundaries, the effectiveness of the surface irrigation system can be improved [4]. To use the cut-off approach, a farmer must make numerous field rounds to determine when the dose of water has reached a particular distance from the starting end or away from the tail end. But this investigation could be difficult, therefore, advanced techniques are required for accurate measurements to cut-off water supply and to save water while increasing crop productivity. A common surface irrigation system is fitted with soil moisture sensors, which may help supply minimal water without over-irrigating [5]. Although there are many different hydrological models, the WinSRFR is a more sophisticated version of the SRFR model and offers more computational possibilities than other hydrological models [6]. In this study, the traditional sensor-based systems were also integrated with Wi-Fi and computer communication networks to automate Pakistan's irrigation system and determine the actual irrigation interval.

2. Materials and Methods

2.1. Experimental Site

This experiment was carried out at the Postgraduate Agriculture Research Station (PARS), University of Agriculture, and Faisalabad-Pakistan. For the experiment, a field of 33.84 m × 27.44 m was divided into 30 equal grids. Each grid was 5.64 m × 5.488 m, and soil samples were collected from each grid at the depth of 9 in for physical and chemical analysis. The field capacity in the experimental field area was determined using a soil moisture tester.

2.2. Hydraulic Simulation Model of Surface Irrigation System

In the present study, the USDA-developed WinSRFR 4.1.3 was used to simulate a surface irrigation system. The WinSRFR is the latest and most popular model being used for this type of study [7].

2.3. Model Calibration

The location of the single directing point affects the flow rate, cut-off time, and infiltration parameters. During land preparation, which includes making a border with varying width and length at the experimental site, 10 pegs of 30 cm height were positioned at equal intervals along the borders. Subsequently, the time was noted when the water reached each peg. Finally, the WinSRFR model and the advance times derived from field data were compared.

2.4. Experimental Design in WinSRFR

Almost ninety treatments were created in the model by combining different border widths, discharge cut-offs, and inflow cut-offs concerning distance. However, all the other parameters including field length, field slope, and field depth, etc. were held constant during the simulation process. Among these total treatments, 19 (one traditional and 18 simulated) were used in the actual field studies. These treatments had three levels for sensor position (55%, 65%, and 75%) and border width (6.4 m, 8.5 m, and 10.7 m).

2.5. Water Use Efficiency

In this study, crop yield and the total water consumed by the crop over the season was used to calculate water use efficiency (Equation (1)), which is also known as true agricultural water productivity [8].

$$\text{CWP} = \frac{\text{Grain yield (Kg ha}^{-1}\text{)}}{\text{Water applied (mm)}} \tag{1}$$

3. Results and Discussion

3.1. Soil Chemical Analysis

The analysis of the soil samples showed that the soil pH ranged from 7.7 to 8.7 and had a range of 2 to 11 ppm for the readily available phosphorus. High phosphorus concentration in soil is recommended [9] because it appears to be crucial for soil fertility and for preventing soil from becoming zinc deficient. Additionally, the range of the amount of accessible nitrogen in soil samples was between 0.017 to 0.042%.

3.2. Moisture Sensors Calibration

The relationship between the moisture content and the sensor reading resistivity was successfully established. These sensors are excellent for irrigation scheduling due to the high coefficient of determination i.e., 0.98 (Figure 1).

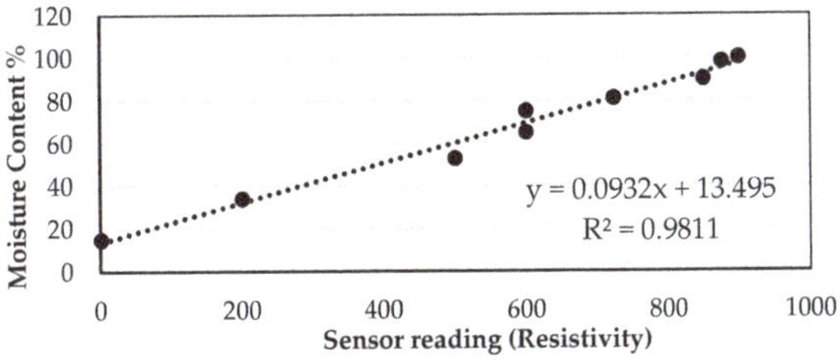

Figure 1. Comparison between soil moisture content (%) and sensor reading (resistivity).

3.3. WinSRFR Hydraulic Simulations

The WinSRFR model was successfully calibrated, and to prevent over- or under-irrigation, hydraulic simulations of all the treatments were run on the model. Application efficiency (AE) alone cannot provide a reliable indicator of how well surface irrigation is working. Additional measurements are needed that include the distribution uniformity lower quarter (DUlq) and the distribution uniformity minimum (DUmin) [10]. These three performance indices (i.e., AE = 92, DUmin = 87, and DUlq = 91) showed higher values, indicating that the treatment T_{10} has excellent uniformity and efficiency.

3.4. Plant Growth Parameters for Maize

The crop yield is affected by various plant growth factors, including the plant population per unit area, the plant height, the number of leaves per plant, the leaf area index, the length of a cob, the number of cobs per row, and the number of seeds per cob. The average number of plants for the treatments T_{10} and T_{11} was 230 and 225, respectively. The lowest counts were in the second year when there was an average of 215 and 170 plants for T_{12} and T_{14}, respectively. The smallest plant height in T_9 was 146.3 cm, while the largest in T_{14} was 181.44 cm.

3.5. Water Use Efficiency

Water use efficiency provides information about how effectively the field's crop utilized the applied water. A sensor was used to compare the differences in the irrigation water application between the two years based on the total rainfall and the soil moisture levels. Figure 2 displays the Water use Efficiency (WUE) of maize for the year 2016–2017 and 2017–2018. The Maximum WUE was 15.78 kg/ha/mm observed in T_{10}; 13.88 kg/ha/mm in T_{11}; 13.07 kg/ha/mm in T_1; and 3.17 kg/ha/mm using the conventional method in year 1. WUE for the treatment T_{10} in the second Year was 14.93 kg/ha/mm, followed by the treatment T_{11} at 13.64 kg/ha/mm and the treatment T_1 at 12.38 kg/ha/mm. The conventional treatment method showed the lowest WUE, at 3.64 kg/ha/mm.

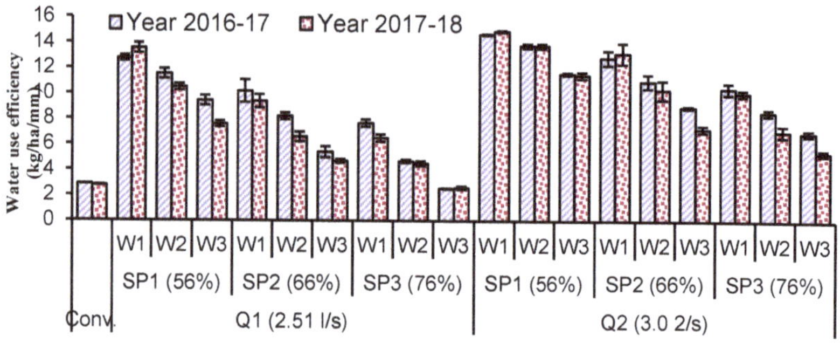

Figure 2. Comparison of WUE of Maize for the Year 2016–2017 and 2017–2018.

4. Conclusions

Managing irrigation practices precisely plays a vital role in saving ample amounts of water with an increase in WUE. The results of the present study revealed that the treatment T_{10} conserved the most water (1121 mm) and produced the most water (15.78 kg/ha/mm), followed by the treatments T_{11}, T_1, and T_{13}, respectively, compared to the control. All the Q_2 treatments (T_{10} to T_{18}) had higher efficiency and uniformity indices values than the corresponding Q_1 treatments (T_1 to T_9). The WUE was decreased by expanding the border because the water did not distribute evenly throughout the area. The reduced border width, ending the irrigation early, and the use of a high inflow rate together resulted in the improved hydraulic performance of surface irrigation, as demonstrated by the efficiency and the uniformity indicators.

Author Contributions: Conceptualization, methodology, software, validation, M.A.A., M.J.M.C., S.S., S.H., and A.B.; formal analysis, S.S., A.B., M.S.W. and M.J.M.C.; investigation, S.S.; data curation, M.A.A., M.J.M.C., S.S., and M.S.W.; writing—original draft preparation, M.A.A.; writing—review and editing, S.S., A.B., S.H., and M.S.W. All authors have read and agreed to the published version of the manuscript.

Funding: This research received no external funding.

Institutional Review Board Statement: Not applicable.

Informed Consent Statement: Not applicable.

Data Availability Statement: Not Applicable.

Conflicts of Interest: The authors declare no conflict of interest.

References

1. Falkenmark, M. The massive water scarcity now threatening Africa: Why isn't it being addressed? *Ambio* **1989**, *18*, 112–118.
2. Alam, S. Globalization, poverty and environmental degradation: Sustainable development in Pakistan. *J. Sustain. Dev.* **2010**, *3*, 103. [CrossRef]
3. Frater, F. Development of an optimisation framework for the investigation of multiple water stores. In *Engineering: Theses and Dissertations*; University of Canterbury: Christchurch, New Zealand, 2022.
4. De Villiers, M. *Water: The Fate of Our most Precious Resource*; Houghton Mifflin Harcourt: Boston, MA, USA, 2001.
5. Wade, R. The system of administrative and political corruption: Canal irrigation in South India. *J. Dev. Stud.* **1982**, *18*, 287–328. [CrossRef]
6. Gillies, M.H.; Foley, J.P.; McCarthy, A.C. Improving surface irrigation. In *Advances in Agricultural Machinery and Technologies*; CRC Press: Boca Raton, FL, USA, 2018; pp. 225–261.
7. Bautista, E.; Schlegel, J.L.; Strelkoff, T.S. *WinSRFR 4.1-User Manual*; USDA-ARS Arid Land Agricultural Research Center: Maricopa, AZ, USA, 2012.
8. Centritto, M.; Loreto, F.; Massacci, A.; Pietrini, F.; Villani, M.C.; Zacchini, M. Improved growth and water use efficiency of cherry saplings under reduced light intensity. *Ecol. Res.* **2000**, *15*, 385–392. [CrossRef]

9. Alloway, B.J. Soil factors associated with zinc deficiency in crops and humans. *Environ. Geochem. Health* **2009**, *31*, 537–548. [CrossRef] [PubMed]
10. Thabet, A. Optimizing the Irrigation Performance of Raised Bed Wheat Using the WinSRFR Model. Master's thesis, Alexandria University, Alexandria, Egypt, 2022. [CrossRef]

environmental sciences proceedings

MDPI

Proceeding Paper

Drone and Robotics Roadmap for Agriculture Crops in Pakistan: A Review †

Ubaid ur Rehman [1,2,*], Tahir Iqbal [1,3], Saddam Hussain [4,5,6], Muhammad Jehanzeb Masud Cheema [4,7], Fahad Iqbal [4] and Abdul Basit [2,7]

1 Department of Farm Machinery and Precision Engineering, PMAS-Arid Agriculture University, Rawalpindi 54000, Pakistan
2 Green AI, Center of Precision Agriculture, PMAS-Arid Agriculture University, Rawalpindi 54000, Pakistan
3 Pilot Project for Data-Driven Smart Decision Platform (DDSDP), PMAS-Arid Agriculture University, Rawalpindi 38000, Pakistan
4 National Center of Industrial Biotechnology (NCIB), PMAS-Arid Agriculture University, Rawalpindi 54000, Pakistan
5 Department of Irrigation and Drainage, University of Agriculture Faisalabad, Faisalabad 38000, Pakistan
6 Department of Agricultural and Biological Engineering, University of California (UC Davis), Davis, CA 95616, USA
7 Faculty of Agricultural Engineering and Technology, PMAS-Arid Agriculture University, Rawalpindi 46000, Pakistan
* Correspondence: ubaid.rehman@uaar.edu.pk; Tel.: +92-332-5650295

† Presented at International Precision Agriculture Pakistan Conference, PMAS-AAUR, Rawalpindi, Pakistan, 22–24 September 2022.

Abstract: Precision agriculture is getting immense attention from researchers and farmers across the world due to the threatening situation of the demand and production gap. Evolution in the electromechanical system and the emergence of intelligent monitoring and conditioning systems have enabled closing the gap to make agronomy quicker, lesser prone to infestations, and still profitable at the same time. Whereas the Internet of Things (IoT) has enabled access to relevant data remotely and automates essential response systems to any threat or requirement by a plant in a particular environment. This study concentrates on gathering such advanced mechatronic techniques in the agricultural sector and analyses of the benefits and disadvantages of the modern method.

Keywords: mechatronics; precision agriculture; IoT; robotics; UAV

Citation: ur Rehman, U.; Iqbal, T.; Hussain, S.; Cheema, M.J.M.; Iqbal, F.; Basit, A. Drone and Robotics Roadmap for Agriculture Crops in Pakistan: A Review. *Environ. Sci. Proc.* **2022**, 23, 39. https://doi.org/10.3390/environsciproc2022023039

Academic Editor: Shoaib Rashid Saleem

Published: 3 February 2023

1. Introduction

The agricultural development [1,2] is the most important factor in Pakistan's economic growth. Around 45% of the labor force and 23.4% of the country's GDP are dependent on agriculture. Numerous constraints and difficulties are faced by local farmers, including the need to increase sustainability, increase yields to satisfy demand, maintain margins, and control water usage. Local farmers usually collect field data manually by monitoring their crops and use the field data to analyze the number of pesticides to be sprayed, soil enrichment chemicals, time to harvest, and yield predictions. Even though this procedure can be as simple as going across a field and making observations, farmers are restricted by time, data recording, and the capabilities of analysis tools while making visual observations and are required to take suitable actions to boost crop yield. Farmers all over the world are employing precision agricultural techniques with the help of aerial/ground robots equipped with sensors/actuators and related decision-making software to overcome problems. These approaches assist farmers at various phases of crop growth.

More than 60% of the people in Pakistan [1–3] are dependent on agriculture for survival. Past supporting an agrarian economy, one of the biggest challenges faced by the agriculture sector is a low per-acre yield due to the lack of using precision agricultural techniques. There is a huge opportunity for innovators and entrepreneurs to bring ideas that can help to integrate information systems, current mobile networks, and sensors/actuators equipped with automated machines and mobile platforms to bring a system that can revolutionize precision agriculture in Pakistan.

Mechatronics engineering [4,5] is the practice of designing electromechanical systems under computer control. Because the mechanical system design must be implemented along with the electrical/electronic and computer control components that will make up the overall system, it might be referred to as "current mechanical engineering design." Examples of mechatronic systems include wired aircraft controls, anti-lock brakes, computer hard drives, CD/DVD players, and videocassette recorders (VCRs) (ABS). These products are all mechanical, but they all rely on integrated electrical and computer control systems to operate.

Electromechanical systems [6] have been used in agriculture for decades. With the advancement of technology, these devices are increasingly being diverted towards mechatronics, as electromechanical equipment is becoming more and more 'intelligent.' In a modern and complex agricultural facility, we can find a lot of embedded systems, such as microcomputers and microcontrollers. These independent parts make up the majority of an agricultural complex in the modern era. Currently, precision agriculture is the cornerstone of modern agriculture (PA). Given instances of mechatronic technologies used in modern agriculture and demonstrates the possibilities of precision farming (such as the Internet of Things). It discusses trends and also defines a vision, as agriculture technologies are in a single network with both pre- and post-technologies.

2. Smart Applications and Robotics

J.B. Grau et al. [7] has done a project with different universities and other agriculture companies to make an optimized system to help with the precision of agriculture with the help of mechatronics. The project's components comprised the analysis and decoding of images of the soil surface, fertility evaluation and the production of a fertility map, signal creation for mechatronic dosing devices, intelligent fertilizer dosing, testing of prototypes, modeling of human-machine interaction, and the development of training materials. This study resulted in the creation of an intelligent and autonomous system that helped identify various soil types and make decisions for the fertilization of the soil under those findings.

Khairul Azmi Mahadhir et al. [8] made an agriculture robot that helped in the identification of terrain conditions. This was a smart robot that learned to adapt to its environment by using Support Vector Machines (SVM), a sort of machine learning. This robot has distinguished between three different types of terrain: sand, gravel, and flora. Figure 1 shows the mechanism of the proposed robot.

Figure 1 shows the hardware and software of the proposed robot, SVM was used to identify terrain by the data given by MEMS sensors vibration. The controller received the identified signal to control the speed of the DC motors and the robot's movement, and if the terrain was sandy, the controller increased the current to the DC motors to provide more torque for the robot's movement.

Slaughter et al. [9] reviewed the four methods used today for autonomous robotic weed management: navigation, detection and identification, precision in-row weed control, and mapping. There was a lot of work done in the fields of guidance, precise row weed management, and mapping. To exploit the weeds for the intended commercial purpose, a lot of work had to be done to identify and discover them.

Saptasagare and Kodada [10] created a sensor and microcontroller system to aid with precision farming. In this scenario, a sensor was used to collect real-time data from the field, such as temperature, water level, and soil moisture. This information was delivered by a microcontroller to the global system for the Mobile communication node. The farmers' computers received the information that had been gathered. This allowed farmers to evaluate the condition of the soil from a distance.

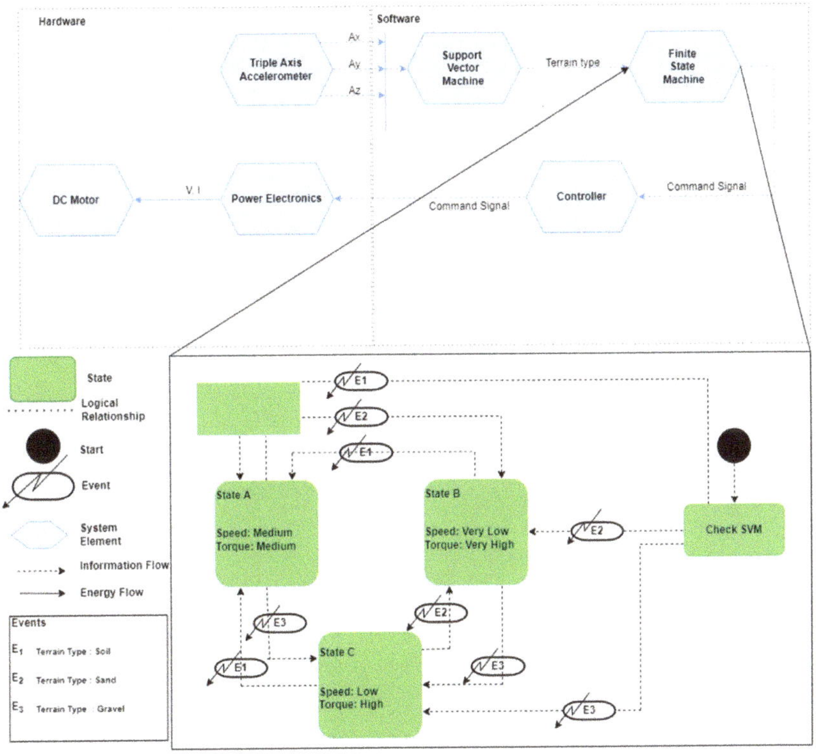

Figure 1. Proposed Robot Mechanism by Mahadhir et al. [8].

Amrutha et al. [11] made a system that took care of soil fertility. Soil fertility depletes during every harvesting of a crop while nutrition, which is needed for the soil, is ignored by the farmer. For this, a system was developed which consisted of a microcontroller, sensor, and mechanical system which gave the needed amount of nutrition automatically after testing the soil. Figure 2 displays the suggested method's flowchart.

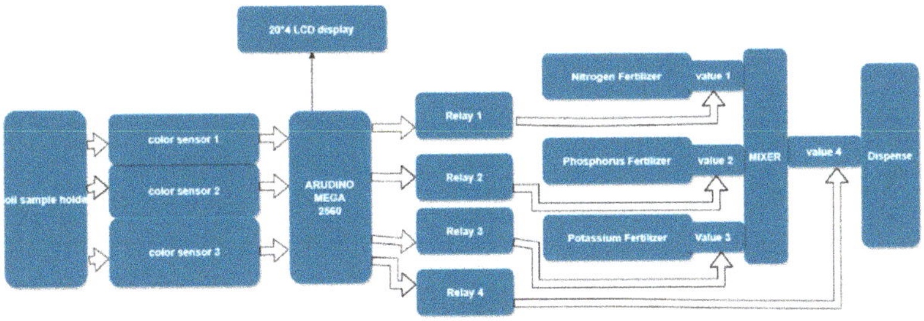

Figure 2. Block representation of the suggested system by Amrutha A et al. [11].

The system consisted of three parts: The first was the testing of the soil. For this purpose, 'colorimetry' was used, on which sensors were used to test the soil color for the needed nutrition. The second was the microcontroller consisting of Adriano mega, which tested the soil and instructed the third section. The third section consists of the mechanic part which took instructions from the microcontroller and opened the valve of the nutrition accordingly.

Kuan-Ming Lin and Chung-Liang Chang [12] offered a concept for a small-scale, intelligent agricultural machine that could perform multiple tasks at once and use artificial vision to automatically weed and regulate the irrigation rate inside a farm. HSV (hue (H), saturation (S), value (V)) color conversion and threshold estimation are two examples of image processing techniques. To confirm the positions of the plants and weeds, the morphological operator's methods and approach were applied. The results are being used to direct watering and weeding operations.

G. Vellidis et al. [13] proposed a method including a smart sensor system for a cotton field to schedule the irrigation system. The system consists of 20 sensors placed in the field and sends data to the circuit board. The circuit board collected the data and sent it to an RFID system which then sent the data to a receiver. Data was then analyzed by the computer software, which determined the amount of water and timing in the cotton field.

Aqeel-ur-Rehman et al. [14] proposed a context-aware sensor grid system for agriculture. This system consists of sensors, temperature, ambient light, and soil moisture probes placed in the field. These sensors give data transmitters to transmit data over satellites, and a grid system receives the data over the network. The system analyzes the data and gives a signal to the actuators to sprinkle or irrigate the field.

Li Qing-Hua et al. [15] gave a system for the tagging of seeds using Radio Frequency Identification (RFID). This system was an implementation of RFID in agricultural seed quality management, covering everything from seed production, storage, and transfer to quality control by relevant enforcement and supervision departments, all the way up to quality assurance and plant information for the end-user, the grower. A seed quality tracking system uses several key strategies, including function design, tag type selection, frequency selection, protocol selection, data security design, anti-collision technology planning, etc. There were several insightful discoveries.

Prathibha S R et al. [16] presented a method employing the Internet of Things (IoT) for precise smart agriculture. The system had a humidity and temperature sensor that transmitted data to a microcontroller unit, which processed and transmitted it to the farmer through Wi-Fi. Additionally, a mounted camera took pictures and relayed them to the farmer after a predetermined amount of time. The internet was used in this fashion to keep the farmer informed about his property while he was away. The block diagram of the IOT-based system is displayed in Figure 3.

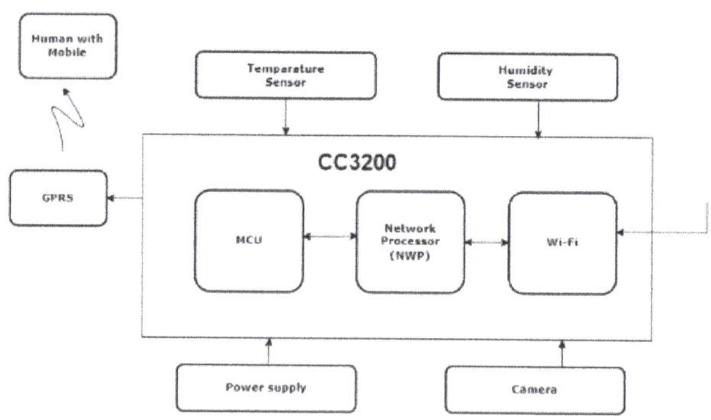

Figure 3. Block diagram of IOT Based System proposed by Prathibha S R et al. [16].

Aarne Halme et al. [17] proposed a robot Work Partner (WP) that was a mobile service robot prototype with a centaur-like look that was made to interact with people outside. It had four wheels and four legs and had a humanoid appearance in the upper torso. The robot was able to complete tasks just as well as a human could. Robots may be used in place of or in addition to humans in the workplace

Arindam Giri [18] suggested developing a system called AgriTech to automate agricultural processes using smart devices, WSN networks, and an Internet device. The farmer might be able to better monitor crops and farmland from a distance if they have a mobile device in hand. A farmer could operate agricultural equipment like an autonomous water sprayer to use in the field of agriculture using smart mobile phones. Therefore, this technology may lessen the need for human labor in the agricultural industry.

Peng Zhang et al. [19] has proposed that IoT and Big Data can be used to irrigate and fertilize agricultural fields. It consists of two parts, and each part has four modules. The first module's data collecting layer collected a sizable amount of data from the farmers and stored it in the system. The intelligent layer stored the data accordingly in the big data center and connected it to the decision layer. The second part is all the field management, and it decides whether the crops need fertilizer or water based on the data stored in the Big Data system and sent to the farmer using the IoT system. Figure 4 shows the structure of the Big Data system.

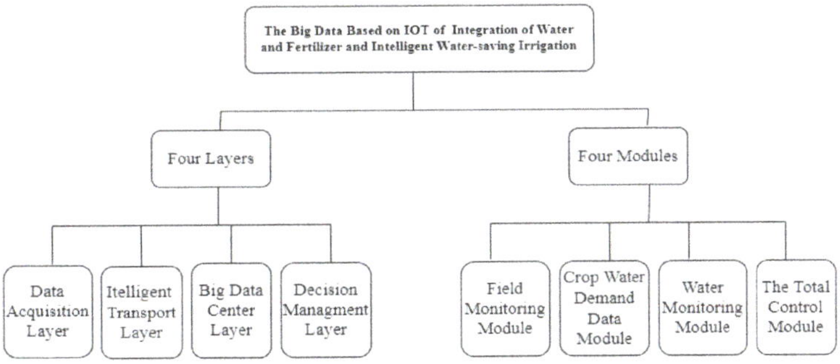

Figure 4. Structure Diagram [19].

Sinung Suakanto et al. [20] and Manlio Bacco [21] proposed an IOT based system for smart farming. The conceptual model and system are designed to assist smart farming decisions, and many are constructed using network sensor applications that execute the tasks that farmers need to be able to accomplish the activities. Agriculture will adopt a comprehensive system based on an Internet of Things (IoT) strategy. Data analysis, task management, control, and sensor-based data collection are the primary areas of emphasis in model creation and system design. This method addresses the issues farmers experience with task management, planning, evaluating environmental conditions, and information exchange.

Min Zhang and Qinggang Meng [22] proposed a detection method of the canker lesion disease citrus canker from field-collected leaf photographs, which were more difficult to obtain than leaf images created in a lab. In order to segregate the lesions from their background using an upgraded AdaBoost algorithm, the most crucial properties of citrus lesions were first determined. Then, a canker lesion descriptor that integrates the color and local texture distribution of canker lesion zones as advised by plant phytopathologists is presented.

Rumpf et al. [23] discussed a method to find the disease in the early stages in sugar beets. After inoculation with the pathogens Cercospora beticola, Uromyces betae, or Erysiphe betae that cause Cercospora leaf spots, sugar beet rust, and powdery mildew, respectively, hyperspectral data were collected from healthy leaves and leaves treated with those pathogens for 21 days. The computerized classification used nine spectrophotometric vegetation indices linked to physiological factors. The support vector machine with a radial basis serving as a nucleus allows for early separation between healthy and infected plants, as well as between particular diseases.

S. Meivel [24] suggested a project that used an unmanned aerial vehicle to monitor and apply urea in the field (UAV). The project offered a high-performance quadrant model with a focus on high execution; it enabled all directions (0°–360°) with large payloads and lighter-weight materials, as well as cutting-edge brushless BLDC motors. The drone had a maximum height of 8 kilos, a diameter of 1.8 m, and a total weight of 3 L. The maximum flight time was about 60 min without the payload. The Arduino was used to remotely program the drone. The RTOS system in the drone provided a solution for every urgent circumstance. The UAV created for the suggested project is shown in Figure 5.

Figure 5. Designed UAV by S. Meivel [24].

3. Conclusions

The modern era is evolving towards smart technology while agronomy, being an integral ecological constituent, is seeing its fair share of such advancements. Mechatronics, being a multi-discipline engineering field, intends to aid the agriculture sector through robots and automation. Concerned literature shows an increasing trend in the usage/prospection of such electromechanical systems that could assist farmers in various aspects of agronomy. From seeding to harvesting, several devised/proposed systems could automate labor-intensive procedures such as implanting the seed, cutting or spraying weeds, and collecting the fruits, crops, and vegetables from the fields with precise actuators controlled by high-end control systems. Such systems make agricultural procedures precise, cost-effective, and more profitable. UAV and on-ground robots have been proposed and have pros and cons of their own, such as on-ground robots being more functional but tricky to operate in humid and rough terrains while UAVs have better reach but limited energy and capabilities.

On the other hand, the studies in this field suggest smart systems for online information collection/access through IoT and smart monitoring systems that use state-of-the-art Machine Learning (ML) and clustering methods. Greenhouse and outdoor temperature, soil humidity, nutrition monitoring, and broadcasting or online data access make it convenient for farmers to retrieve vital information remotely in real-time. Although the proposed ML methods for detecting/segregating diseases, organs, and weeds are fairly accurate but need more maturity for being robust and readily adaptive for any kind of terrain and plants.

Mechatronics could help the industry in the following methods:

1. The use of automation and intelligent systems in irrigation will help to improve irrigation efficiency and lower water waste.
2. The use of automation and intelligent systems in the food industry also aids in determining the best chemicals and pesticides to use for agricultural purposes.
3. Automatic and intelligent systems help pick out and manage weeds. The multitasking robots will support agricultural activities and processes, as well as finish the task quickly, preserve the product's quality, and reduce the need for human effort.
4. We will be able to supply information on the humidity, temperature, and water level thanks to the hybrid agricultural systems.
5. We can streamline the agricultural process and choose the best weed, crop, and pesticides by better utilizing automation and IoT in the sector of agriculture.

6. On the other side, automation and IoT systems offer solutions and make the agricultural process more predictable.
7. These technologies will contribute to reducing human labor requirements and raising production.

Author Contributions: Conceptualization, methodology, software, validation, U.u.R., T.I., S.H. and A.B.; formal analysis, M.J.M.C.; investigation, F.I.; resources, U.u.R. and S.H.; data curation, U.u.R. and T.I.; writing—original draft preparation, U.u.R.; writing—review and editing, A.B. and S.H.; visualization, M.J.M.C. All authors have read and agreed to the published version of the manuscript.

Funding: This research received no external funding.

Institutional Review Board Statement: Not applicable.

Informed Consent Statement: Not applicable.

Data Availability Statement: The data can be available on request.

Conflicts of Interest: The authors declare no conflict of interest.

References

1. Azam, A.; Shafique, M. Agriculture in Pakistan and its Impact on Economy—A Review. *Int. J. Adv. Sci. Technol.* **2017**, *103*, 47–60. [CrossRef]
2. Waqas, M.S.; Cheema, M.J.M.; Hussain, S.; Ullah, M.K.; Iqbal, M.M. Delayed irrigation: An approach to enhance crop water productivity and to investigate its effects on potato yield and growth parameters. *Agric. Water Manag.* **2021**, *245*, 106576. [CrossRef]
3. Malik, S.; Hussain, S.; Waqas, M.S. Effect of water quality and different meals on growth of *Catla catla* and *Labeo rohita*. *Big Data Water Resour. Eng.* **2020**, *1*, 4–8. [CrossRef]
4. Altintas, Y.; Croft, E.A. Electro-mechanical design engineering: A progress report and future directions for mechatronics education. *Int. J. Mech. Eng. Educ.* **2002**, *30*, 325–339. [CrossRef]
5. Waqas, M.S.; Cheema, M.J.M.; Waqas, A.; Hussain, S. Enhancing water productivity of potato (*Solanum tuberosum* L.) through drip irrigation system. In Proceedings of the 2nd International Conference on Horticultural Sciences, Faisalabad, Pakistan, 18–20 February 2016; pp. 18–20.
6. Kuru, K.; Yetgin, H. Transformation to advanced mechatronics systems within new industrial revolution: A novel framework in automation of everything (AoE). *IEEE Access* **2019**, *7*, 41395–41415. [CrossRef]
7. Grau, J.; Anton, J.; Packianather, M.; Ermolov, I.; Aphanasiev, R.; Cisneros, J.; Cortina-Januchs, M.; Jevtić, A.; Andina, D. Sustainable agriculture using an intelligent mechatronic system. In Proceedings of the 2009 35th Annual Conference of IEEE Industrial Electronics, Porto, Portugal, 3–5 November 2009; pp. 3416–3421.
8. Mahadhir, K.A.; Tan, S.C.; Low, C.Y.; Dumitrescu, R.; Amin, A.T.M.; Jaffar, A. Terrain classification for track-driven agricultural robots. *Procedia Technol.* **2014**, *15*, 775–782. [CrossRef]
9. Slaughter, D.C.; Giles, D.; Downey, D. Autonomous robotic weed control systems: A review. *Comput. Electron. Agric.* **2008**, *61*, 63–78. [CrossRef]
10. Saptasagare, V.S.; Kodada, B.B. Real-time implementation and analysis of crop-field for agriculture management system based on microcontroller with GPRS (M-GPRS) and SMS. *Int. J. Comput. Appl.* **2014**, *98*, 1–6. [CrossRef]
11. Amrutha, A.; Lekha, R.; Sreedevi, A. Automatic soil nutrient detection and fertilizer dispensary system. In Proceedings of the 2016 International Conference on Robotics: Current Trends and Future Challenges (RCTFC), Thanjavur, India, 19–20 December 2016; pp. 1–5.
12. Chang, C.-L.; Lin, K.-M. Smart agricultural machine with a computer vision-based weeding and variable-rate irrigation scheme. *Robotics* **2018**, *7*, 38. [CrossRef]
13. Vellidis, G.; Tucker, M.; Perry, C.; Kvien, C.; Bednarz, C. A real-time wireless smart sensor array for scheduling irrigation. *Comput. Electron. Agric.* **2008**, *61*, 44–50. [CrossRef]
14. Aqeel-ur-Rehman, Z.A.S.; Shaikh, N.A.; Islam, N. An integrated framework to develop context-aware sensor grid for agriculture. *Aust. J. Basic Appl. Sci.* **2010**, *4*, 922–931.
15. Li, H.; Wang, H.-j.; Shang, Z.; Li, Q.-h.; Xiao, W. Application of RFID in agricultural seed quality tracking system. In Proceedings of the 2010 8th World Congress on Intelligent Control and Automation, Jinan, China, 7–9 July 2010; pp. 3073–3077.
16. Prathibha, S.; Hongal, A.; Jyothi, M. IoT based monitoring system in smart agriculture. In Proceedings of the 2017 International Conference on Recent Advances in Electronics and Communication Technology (ICRAECT), Bangalore, India, 16-17 March 2017; pp. 81–84.
17. Halme, A.; Luksch, T.; Ylönen, S. Biomimicing motion control of the WorkPartner robot. *Ind. Robot Int. J.* **2004**, *31*, 209–217. [CrossRef]

18. Giri, A.; Dutta, S.; Neogy, S. Enabling agricultural automation to optimize utilization of water, fertilizer and insecticides by implementing Internet of Things (IoT). In Proceedings of the 2016 International Conference on Information Technology (InCITe)—The Next Generation IT Summit on the Theme-Internet of Things: Connect your Worlds, Noida, India, 6–7 October 2016; pp. 125–131.
19. Zhang, P.; Zhang, Q.; Liu, F.; Li, J.; Cao, N.; Song, C. The construction of the integration of water and fertilizer smart water saving irrigation system based on big data. In Proceedings of the 2017 IEEE International Conference on Computational Science and Engineering (CSE) and IEEE International Conference on Embedded and Ubiquitous Computing (EUC), Guangzhou, China, 21–24 July 2017; pp. 392–397.
20. Suakanto, S.; Engel, V.J.; Hutagalung, M.; Angela, D. Sensor networks data acquisition and task management for decision support of smart farming. In Proceedings of the 2016 International Conference on Information Technology Systems and Innovation (ICITSI), Bandung, Indonesia, 24–27 October 2016; pp. 1–5.
21. Bacco, M.; Berton, A.; Ferro, E.; Gennaro, C.; Gotta, A.; Matteoli, S.; Paonessa, F.; Ruggeri, M.; Virone, G.; Zanella, A. Smart farming: Opportunities, challenges and technology enablers. In Proceedings of the 2018 IoT Vertical and Topical Summit on Agriculture-Tuscany (IOT Tuscany), Tuscany, Italy, 8–9 May 2018; pp. 1–6.
22. Zhang, M.; Meng, Q. Automatic citrus canker detection from leaf images captured in field. *Pattern Recognit. Lett.* **2011**, *32*, 2036–2046. [CrossRef]
23. Rumpf, T.; Mahlein, A.-K.; Steiner, U.; Oerke, E.-C.; Dehne, H.-W.; Plümer, L. Early detection and classification of plant diseases with support vector machines based on hyperspectral reflectance. *Comput. Electron. Agric.* **2010**, *74*, 91–99. [CrossRef]
24. Meivel, S.; Dinakaran, K.; Gandhiraj, N.; Srinivasan, M. Remote sensing for UREA Spraying Agricultural (UAV) system. In Proceedings of the 2016 3rd International Conference on Advanced Computing and Communication Systems (ICACCS), Coimbatore, India, 22–23 January 2016; pp. 1–6.

environmental sciences proceedings

MDPI

Proceeding Paper

Downscaling of Satellite Rainfall Data Using Remotely Sensed NDVI and Topographic Datasets †

Zarina Yasmeen [1,*], Muhammad Jehanzeb Masud Cheema [2,3], Saddam Hussain [3,4,5], Zainab Haroon [2], Sadaf Amin [6] and Muhammad Sohail Waqas [6]

1 Agricultural Mechanization Research Institute (AMRI), Agricultural Department (Field Wing), Government of Punjab, Lahore 53400, Pakistan
2 Faculty of Agricultural Engineering and Technology, PMAS-Arid Agriculture University, Rawalpindi 46000, Pakistan
3 National Center for Industrial Biotechnology, PMAS-Arid Agriculture University, Rawalpindi 46000, Pakistan
4 Department of Irrigation and Drainage, University of Agriculture Faisalabad, Faisalabad 38000, Pakistan
5 Department of Agricultural and Biological Engineering, University of California (UC Davis), Davis, CA 95616, USA
6 Soil Conservation Group, Agriculture Department (Field Wing), Government of the Punjab, Rawalpindi 46000, Pakistan
* Correspondence: zarinayasmeen4@gmail.com

† Presented at the 1st International Precision Agriculture Pakistan Conference 2022 (PAPC 2022)—Change the Culture of Agriculture, Rawalpindi, Pakistan, 22–24 September 2022.

Abstract: Rainfall is a key factor in hydrological, meteorological, and water management applications in restricted regions or basins, but its measurement remains difficult in mountainous or otherwise remote places due to a lack of readily available rain gauges. While satellite rainfall data offer a better temporal resolution than other sources, the majority of this data are only available at a coarse geographic resolution, which distorts the true picture of precipitation. Thus, researchers at the University of Agriculture in Faisalabad used the normalized difference vegetation index (NDVI) monthly data and 1 km topography data for the whole Indus Basin from 2002 to 2011 to reduce the TRMM's spatial resolution from 25 km to 1 km. An approach to downscaling based on a regression model with residual correction was established in this study. First, we resampled the NDVI and TRMM datasets to a 25 km resolution and established a regression model connecting the two datasets. Precipitation was forecasted at a distance of 25 km. The TRMM 3B43 product was then adjusted downward by the projected precipitation to achieve the residual value. The IDW method was used to reduce the resolution of the residual image from 25 km to 1 km. Rainfall was predicted using a regression model applied to NDVI at a 1 km spatial resolution. The final downscaled precipitation was created by combining the modeled precipitation at 1 km resolution with the residual image. The result was double-checked by the post-processing steps of validation and calibration.

Keywords: downscaling; rainfall; NDVI; TRMM; spatial resolution; meteorological observation

Citation: Yasmeen, Z.; Cheema, M.J.M.; Hussain, S.; Haroon, Z.; Amin, S.; Waqas, M.S. Downscaling of Satellite Rainfall Data Using Remotely Sensed NDVI and Topographic Datasets. *Environ. Sci. Proc.* **2022**, *23*, 40. https://doi.org/10.3390/environsciproc2022023040

Academic Editor: Muhammad Naveed Anjum

Published: 6 February 2023

1. Introduction

Rainfall is an important factor for hydrological, meteorological, and water management applications in specific regions as well as in basins. Rainfall has a significant impact on agriculture production because all plants require water to survive. A healthy crop needs a consistent pattern of rainfall; an excess or a deficiency of rain can be damaging or even stressful to the crop. Additionally, rainfall is required for other production activities, including irrigation and disaster mitigation [1]. Point rain gauge stations are the standard method for collecting rainfall data. However, accurate basin-level rainfall estimation remains an enormous challenge [2]. More precise rainfall measurements are required for basin-scale water management planning, water accounting, and hydrological research [3–5]. An insufficient number of rain gauges in the Indus Basin makes it difficult to conduct research and provide timely flood warnings, as point measurements do not provide an accurate representation of real average rainfall values; moreover, rain gauges are prone to a variety of random and systematic errors [6]. To address this problem, researchers have created a new downscaling technique that makes use of the correlation between high-resolution TRMM and normalized difference vegetation index (NDVI). This downscaling technique makes use of TRMM precipitation, which estimates the coarse scale, and NDVI patterns on the fine scale. Significant improvements in bias, correlation, and root-mean-square error were found after shrinking the full-period mean annual precipitation for both wet and dry seasons. A downscaling method was used because the 0.25 km resolution is too low for use at the basin scale and a higher resolution is preferred for hydrological studies. For this reason, using 1 km downscaling is advantageous in hydrological applications [7]. The majority of the random errors in rain gauge readings come from two sources: observational errors and instrumental errors, and together they can account for a discrepancy of up to 30% [8].

It is for this reason that satellite-based sensors are needed to provide more precise spatially distributed rainfall estimates [9]. Due to the errors and limitations of rain gauges, remote sensing data is often seen as the best option. Since the development of remote sensing technology, using satellites to estimate rainfall has become a realistic option until proper ground trothing occurs [5]. Space-based sensors cover a wide area and operate for extended periods of time, making it possible for new technologies to significantly improve the accuracy and timeliness with which we estimate rainfall [7]. When satellite rainfall estimates are calibrated and validated correctly, we can gain a deeper understanding of spatially distributed precipitation in both subtropical and tropical regions, while downscaled precipitation estimates can be later validated with a separate precipitation dataset [7].

In this study, we used NDVI monthly data and topography data at a scale of 1 km from 2002 to 2011 for the entire study area to reduce the Tropical Rainfall Measuring Mission's (TRMM) spatial resolution from the value of 25 km.

2. Materials and Methods

The Indus Basin is the study area, and the elevations examined span from zero to 8,616 m above sea level. Pakistan experiences two distinct seasonal patterns across its four distinct climate seasons (summer, winter, spring, and fall). Pakistan spans the coordinates 23°45′–36°75′ North and 61°30′–75°50′ East. Plains, hills, desert, forests, and plateaus can be found in the southern part of the study area, while the Karakoram Mountains can be found in the northern part.

2.1. Data

2.1.1. Tropical Rainfall Measuring Mission

TRMM is a combined collaboration between the National Aeronautics and Space Administration (NASA) and Japan Aerospace Exploration Agency [10]. The NASA Goddard Earth sciences data from 2002 to 2011 was used to obtain the RMM 3B43 data (http://mirador.gsfc.nasa.gov/ (accessed on 1 January 2015)).

2.1.2. Rain Gauge Stations

The Pakistan Metrology Department provided information from their network of rain gauges. Data on rainfall over a specific time was used for this analysis.

2.1.3. NDVI

NDVI is an essential parameter for estimating rainfall and an important index that describes the degree of greenness of land surfaces. Given MODIS' superior spatial resolution and global coverage, we decided to use its NDVI data. The NDVI-containing MODIS data from 2002 to 2011 were obtained from the EOS Gateway (https://lpdaac.usgs.gov/ (accessed on 20 January 2015)).

2.1.4. Digital Elevation Model

The digital elevation model (DEM) 2002–2011 was downloaded from the NASA shuttle radar topography mission website (http://www.glcf.umd.edu/data/srtm (accessed on 30 January 2015)) projected onto the study area by using ERDAS software.

2.2. *Methodology*

The stepwise procedure to downscale precipitation is explained below.

2.2.1. Factor Analysis

The following relationships were determined:
The relationship between NDVI and TRMM annual precipitation.
The relationship between altitude and TRMM annual precipitation.

2.2.2. Establishment of Model

The model was established for regression analysis at a low resolution between NDVI and TRMM, and the precipitation was estimated at 25 km. Mathematically, this can be expressed as shown in (1):

$$\mathrm{f(NDVI,\ DEM) = PTLR} \tag{1}$$

Then, using Equation (1), the relationship with precipitation was estimated at high resolution (1 km). Mathematically, it is expressed as shown in (2):

$$\mathrm{f(NDVI,\ DEM) = PTHR} \tag{2}$$

2.2.3. Residual Conversion

The difference between precipitation and TRMM 3B43 values are residual values (ΔT) at 25 km. By using Inverse Distant Weighting (IDW) interpolation techniques, 25 km residual values were converted to 1 km values by using (3):

$$\mathrm{TRMM - PTLR = \Delta T} \tag{3}$$

2.2.4. Accurate Rainfall

The converted (ΔT) was added to the model, where the NDVI and TRMM data sets were used, and accurate precipitation was calculated by using (4):

$$\mathrm{PTcorr. = PTHR + \Delta THR} \tag{4}$$

2.2.5. Monthly Precipitation from Yearly Precipitation

Monthly precipitation was extracted from annual precipitation by determining the monthly fraction and multiplying it by annual downscale precipitation by using (5) and (6):

$$\mathrm{fraction\ i} = \frac{\mathrm{org.TRMM\ i}}{\sum_{\mathrm{i=1}}^{12} \mathrm{TRMM\ i}} \tag{5}$$

$$\mathrm{PTi = fraction\ i\ \times PTHR} \tag{6}$$

where i in (5) represents monthly precipitation as estimated from the original TRMM 3B43 product.

2.2.6. Validation and Calibration

The downscaled precipitation was validated by using rain gauge data. The slope (b) of the regression analyses and coefficient of determination (R^2), the linear coefficient through the origin (a = 0), and the relative root mean square error (RRMSE) were calculated by using (7), and the root mean square error (RMSE) by using (8):

$$\text{RRMSE} = \sqrt{\frac{\sum_{i=1}^{n}(\text{Oi} - \text{pi})^2}{n}}.100/\text{Oi} \quad (7)$$

$$\text{RMSE} = \sqrt{\frac{\sum_{i=1}^{n}(\text{Oi} - \text{pi})^2}{n}} \quad (8)$$

where P is the yearly precipitation downscaled from TRMM 3B43, O is the observed precipitation from rain gauge stations, Oi is the mean value of the annual/monthly observed values, I is the station number index, and n is the total number of rain gauge stations. The downscaled results were compared to the rain gauge data first, and then the coefficient of determination (R^2) was determined. If the station data and precipitation product have no correlation, the value will be 0, and vice versa if the correlation is perfect. Second, precision was quantified via b. This shows by how much the actual value is higher or lower than what was observed. Third, the root-mean-squared error (RRMSE) gives a percentage value for the dissimilarity between the estimated and observed values. If the RRMSE is less than 10%, the validation is considered excellent; if it is between 10% and 20%, the validation is good; if it is between 20% and 30%, the validation is acceptable; and if it is more than 30%, the validation is poor [11]. By focusing on highly biased results, the RRMSE test gives a measure of the prediction model error rate. Lastly, when it comes to root-mean-square error (RMSE), smaller values are preferable.

3. Results and Discussion

3.1. NDVI and TRMM Regression

Images from the Tropical Rainfall Survey show that rainfall in the northern and north-western parts along the Indus Basin is actually much higher, around 1200 mm y^{-1}, than in the central and southeastern regions, around 400 mm y^{-1}. The normalized differential vegetation index essentially exhibited the same example as TRMM precipitation, and an unmistakable spatial resolution connection exists between TRMM and NDVI as shown in Figure 1. The resolution dependence of the normalized distinct vegetation index and the tropical precipitation measurement duty relationship were tested at different resolutions.

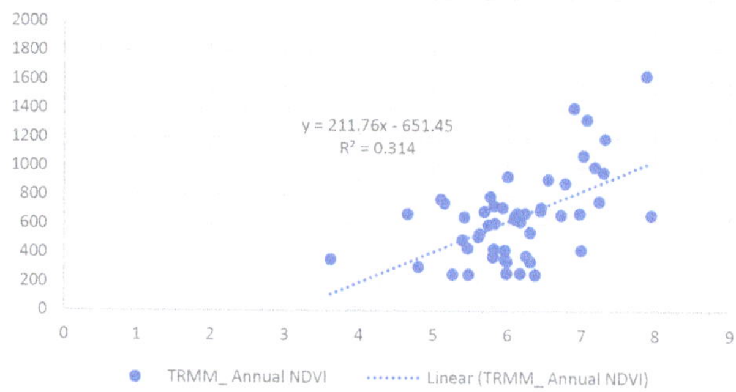

Figure 1. Relationship between NDVI and TRMM.

A consistent exponential relationship at 1 km resolution was demonstrated at these resolutions, showing that an unmistakable relationship between blueness and precipitation exists. The curve shows that the linkages acquired inordinately high NDVI values when the association with dispersed precipitation and water no longer restricted growth. The strongest associations were found in the normalized difference in vegetation index, which varied somewhere between 0.2–0.7, and in precipitation, which varied from 200 to

800 mmy^{-1}. Different functions were tried at a nominal resolution of 25 km, but resulted in a lower R2 value. For the linear function, it was 0.55; for exponential, it was 0.52; and the quadratic polynomial, c was 0.53. The value of the correlation varied with the resolution. The difference vegetation index was normalized in all cases affected by variables, e.g., topography, soil, vegetation type, temperature, irrigation system, and human impact. Using this ideal fit to a smaller degree of determination, we evaluated the neighborhood's precipitation pending a comparative response at the ideal scale. In other words, the degree of greenness, as reflected by the normalized difference vegetation index, is a factor of many different variables, including precipitation.

3.2. Residual Correction

However, after subtracting the modelled TRMM at 25 km from the 3B43 precipitation, several regions had considerable residuals:

$$\text{TRMM} - \text{PTLR} = \Delta\text{T} \tag{9}$$

The other map shows a portion of the precipitation that was not elucidated using only the normalized differential vegetation index. Residues with negative values showed greener areas than expected, specifying rainfall. These negative ranges can have additional water sources (i green has a deep root system). Residual areas with positive values (green) indicate less green than the indicated observed normal precipitation. The abrupt, vegetative low slopes of the sections or ranges of hills with heavy rainfall located in the oversaturation zone of the TRMM-adjusted and normalized Difference Vegetation Index may clarify this residual value. The number of residuals were reduced using the introduction through the internal functions of TRMM (ΔTRMMLR). High-resolution residues were obtained from TRMM using the IDW technique.

3.3. Accurate Precipitation

Final downscaled rainfall was obtained by adding model rainfall at 1 km and residual image at 1 km resolution. The resultant image is shown in Figure 2.

$$\text{PTcorr.} = \text{PTHR} + \Delta\text{THR} \tag{10}$$

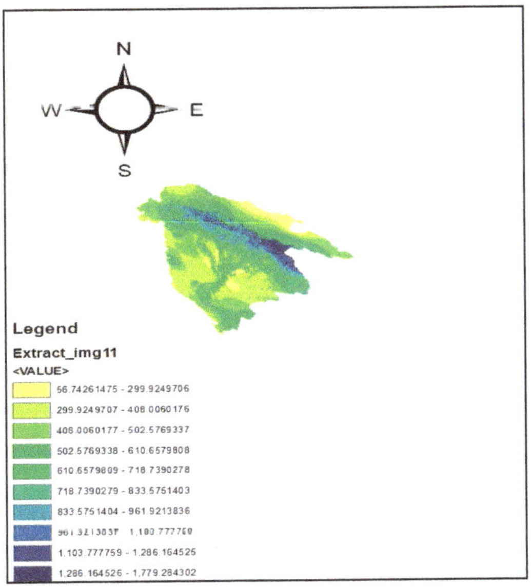

Figure 2. TRMM rainfall at 1 km resolution.

3.4. Validation

Using information from rain gauges, the validity of the proposed method to improve rainfall mapping was accepted. A total of 115 stations across the Indus Basin have complete precipitation records between 2002 and 2011. These stations' data were used in the validation process. Validation indicators were determined after data from the first low-resolution TRMM and the downscaled precipitation value were extracted from the station locations. As shown in Table 1 and Figure 3, the results of the validation showed that the R^2, the bias, and relative root mean square error were significantly improved after applying the downscaling methodology. This downscaling methodology is very promising, and the resulting high-resolution rainfall map was more accurate than the initial TRMM assessment.

Table 1. Original TRMM and final TRMM.

Original TRMM 3B43					TRMM Final			
Year	Original R^2	B	RRMSE%	RMSE mm	Final R^2	b	RRMSE%	RMSE, mm
2002	0.77	0.95	27.84	71.84	0.81	1.17	19.42	50.11
2003	0.77	0.93	28.83	72.69	0.85	1	16.72	42.16
2004	0.78	0.97	23.19	66.98	0.81	1.07	17.63	50.92
2005	0.82	0.99	23.72	81.46	0.87	1.04	14.45	49.62
2006	0.79	0.87	28.31	79.63	0.82	1.12	8.73	24.56
2007	0.78	0.91	33.44	98.57	0.84	1.09	19.15	56.54
2008	0.75	0.94	32.36	86.01	0.77	1.09	21.7	57.68
2009	0.79	1	25.48	75.4	0.84	1.07	16.07	47.58
2010	0.83	0.97	22.78	67.67	0.87	1.09	23.33	69.3
2011	0.77	0.93	30.66	83.99	0.78	1.17	25.81	70.7
Average	0.78	0.94	27.66	78.42	0.82	1.09	18.3	51.91

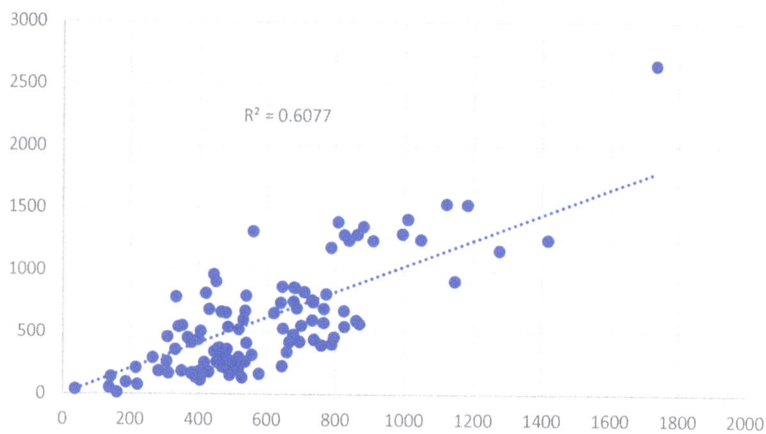

Figure 3. Correlation between downscaled TRMM base on NDVI and rain gauges.

4. Conclusions

Downscaling of the TRMM rain gauge in the Indus River basin was investigated using TRMM 3B43 monthly precipitation products and the NDVI imaging time series from 2002 to 2011. The results of the downscaling method based on a direct scale-dependent relationship between precipitation and NDVI was validated on the precipitation data set. Several conclusions can be drawn from this study:

- This study showed that adopting the degree of vegetation as a substitute for rainfall on an annual basis could boost the geographical resolution and precision of rainfall evaluations.

- The downscaling approach resulted in extremely significant improvements in the accuracy and spatial resolution of the average rainfall from 2003–2011.
- Expanding this method at a higher temporal resolution should be the focus of future research (e.g., regular or month to month).

The primary finding of this study is that the NDVI can be used to improve the spatial resolution of rainfall data from the Tropical Rainfall Measuring Mission (TRMM) in the Indus Basin, and that the approach shown here is universally applicable to other semiarid and dry parts of the world.

5. Recommendation

Further downscaling methods that consider both NDVI and DEM at different spatial (local, national, and global) and higher temporal (weekly and daily) resolutions should be developed in the future. Therefore, it is important to describe the relationship between NDVI and TRMM at various scales (i.e., 75-100 km) and determine the optimal resolution for establishing this relationship. The presented methodology is general in nature and applicable to other semi-arid regions of the world, and in the Indus Basin NDVI can also be used to accurately downscale TRMM precipitation.

Author Contributions: Conceptualization, Z.Y. and M.J.M.C.; methodology, Z.Y.; software, M.J.M.C.; validation, Z.Y., S.H. and M.S.W.; formal analysis, Z.Y. and S.H.; investigation, Z.H. and M.S.W.; resources, M.J.M.C. and S.H.; data curation, Z.Y. and S.A.; writing—original draft preparation, Z.Y., S.H., M.S.W. and S.A.; review and editing, M.J.M.C., S.H. and Z.H.; visualization, S.H.; supervision, M.J.M.C.; project administration, M.J.M.C.; funding acquisition, Z.Y. and M.S.W. All authors have read and agreed to the published version of the manuscript.

Funding: This research received no external funding.

Institutional Review Board Statement: Not applicable.

Informed Consent Statement: Not applicable.

Data Availability Statement: Not applicable.

Conflicts of Interest: The authors declare no conflict of interest.

References

1. Langella, G.; Basile, A.; Bonfante, A.; Terribile, F. High-resolution space–time rainfall analysis using integrated ANN inference systems. *J. Hydrol. (Amst.)* **2010**, *387*, 328–342. [CrossRef]
2. Sawunyama, T.; Sa, D.H.-W. Application of satellite-derived rainfall estimates to extend water resource simulation modelling in South Africa. *Water Sa* **2008**, *34*, 1–10. [CrossRef]
3. Verdin, J.; Klaver, R. Grid-cell-based crop water accounting for the famine early warning system. *Hydrol. Process.* **2002**, *16*, 1617–1630. [CrossRef]
4. Tobin, K.J.; Bennett, M.E. Adjusting satellite precipitation data to facilitate hydrologic modeling. *J. Hydrometeorol.* **2010**, *11*, 966–978. [CrossRef]
5. Cheema, M.J.M.; Bastiaanssen, W.G. Local calibration of remotely sensed rainfall from the TRMM satellite for different periods and spatial scales in the Indus Basin. *Int. J. Remote Sens.* **2012**, *33*, 2603–2627. [CrossRef]
6. Draper, C.S.; Walker, J.P.; Steinle, P.J.; De Jeu, R.A.; Holmes, T.R. An evaluation of AMSR–E derived soil moisture over Australia. *Remote Sens. Environ.* **2009**, *113*, 703–710. [CrossRef]
7. Immerzeel, W.W.; Rutten, M.M.; Droogers, P. Spatial downscaling of TRMM precipitation using vegetative response on the Iberian Peninsula. *Remote Sens. Environ.* **2009**, *113*, 362–370. [CrossRef]
8. WMO. *Guide to Meteorological Instruments and Methods of Observation*; WMO: Geneva, Switzerland, 1996.
9. Huffman, G.J.; Adler, R.F.; Morrissey, M.M.; Bolvin, D.T.; Curtis, S.; Joyce, R.; McGavock, B.; Susskind, J. Global precipitation at one-degree daily resolution from multisatellite observations. *J. Hydrometeorol.* **2001**, *2*, 36–50. [CrossRef]

10. Jia, S.; Zhu, W.; Lű, A.; Yan, T. A statistical spatial downscaling algorithm of TRMM precipitation based on NDVI and DEM in the Qaidam Basin of China. *Remote Sens. Environ.* **2011**, *115*, 3069–3079. [CrossRef]
11. Campi, P.; Palumbo, A.D.; Mastrorilli, M. Evapotranspiration estimation of crops protected by windbreak in a Mediterranean region. *Agric. Water Manag.* **2012**, *104*, 153–162. [CrossRef]

MDPI
St. Alban-Anlage 66
4052 Basel
Switzerland
Tel. +41 61 683 77 34
Fax +41 61 302 89 18
www.mdpi.com

Environmental Sciences Proceedings Editorial Office
E-mail: environsciproc@mdpi.com
www.mdpi.com/journal/environsciproc

www.ingramcontent.com/pod-product-compliance
Lightning Source LLC
Chambersburg PA
CBHW061443210926
51850CB00008B/1591
9783036572147